CARDIOLOGY RESEARCH AND CLINICAL DEVELOPMENTS

TISSUE ENGINEERING OF THE AORTIC HEART VALVE

FUNDAMENTALS AND DEVELOPMENTS

CARDIOLOGY RESEARCH AND CLINICAL DEVELOPMENTS

Additional books in this series can be found on Nova's website under the Series tab.

Additional E-books in this series can be found on Nova's website under the E-book tab.

CARDIOLOGY RESEARCH AND CLINICAL DEVELOPMENTS

TISSUE ENGINEERING OF THE AORTIC HEART VALVE

FUNDAMENTALS AND DEVELOPMENTS

YOSRY S. MORSI

Nova Science Publishers, Inc.
New York

Copyright © 2012 by Nova Science Publishers, Inc.

All rights reserved. No part of this book may be reproduced, stored in a retrieval system or transmitted in any form or by any means: electronic, electrostatic, magnetic, tape, mechanical photocopying, recording or otherwise without the written permission of the Publisher.

For permission to use material from this book please contact us:
Telephone 631-231-7269; Fax 631-231-8175
Web Site: http://www.novapublishers.com

NOTICE TO THE READER

The Publisher has taken reasonable care in the preparation of this book, but makes no expressed or implied warranty of any kind and assumes no responsibility for any errors or omissions. No liability is assumed for incidental or consequential damages in connection with or arising out of information contained in this book. The Publisher shall not be liable for any special, consequential, or exemplary damages resulting, in whole or in part, from the readers' use of, or reliance upon, this material. Any parts of this book based on government reports are so indicated and copyright is claimed for those parts to the extent applicable to compilations of such works.

Independent verification should be sought for any data, advice or recommendations contained in this book. In addition, no responsibility is assumed by the publisher for any injury and/or damage to persons or property arising from any methods, products, instructions, ideas or otherwise contained in this publication.

This publication is designed to provide accurate and authoritative information with regard to the subject matter covered herein. It is sold with the clear understanding that the Publisher is not engaged in rendering legal or any other professional services. If legal or any other expert assistance is required, the services of a competent person should be sought. FROM A DECLARATION OF PARTICIPANTS JOINTLY ADOPTED BY A COMMITTEE OF THE AMERICAN BAR ASSOCIATION AND A COMMITTEE OF PUBLISHERS.

Additional color graphics may be available in the e-book version of this book.

Library of Congress Cataloging-in-Publication Data

Library of Congress Control Number: 2012930424

ISBN: 978-1-61942-939-0

Published by Nova Science Publishers, Inc. ✝ *New York*

This Book is dedicated to

My lovely children, Yasser, Amanda and Adam, and my late mother

Contents

Preface		**ix**
Chapter I	Why Tissue Engineering of Heart Valve	**1**
Chapter II	Basic Introduction to Heart Valve Diseases	**7**
Chapter III	Heart Valve Replacements	**25**
Chapter IV	Biomaterials Characterisation	**43**
Chapter V	Scaffold Fabrication Techniques	**77**
Chapter VI	In Vitro Conditioning— Bioreactors and ECM Generation	**97**
Chapter VII	Concept and Development of Tissue Engineering Aortic Heart Valve	**119**
Chapter VIII	Closing Remarks and Future Challenges	**161**
Index		**179**

Preface

In the last 40 years, the study of aortic heart valve replacements has had its roots in the work of great clinicians, scientists and biomedical engineers who established the techniques for heart valve substitutes. Recently, the research on the development of aortic heart valve replacements has shifted toward tissue engineering (TE), a multidisciplinary research that involves biological, biomaterial and bioengineering sciences. This book introduces current valve substitutes and their limitations and the futuristic tissue engineering of aortic heart valves (TEHV). It highlights the substantial achievements that have been accomplished in the last two decades in TEHV research and discusses the much needed exertion in defining the functional requirements of clinical implementation and in developing the means to produce a functional aortic valve.

Chapter I

Why Tissue Engineering of Heart Valve

1.1. Overview

Healthy biological heart valves are created to permit unidirectional blood flow without regurgitation or trauma to the blood. They do not permit any excessive developments of local stress in the valve leaflets and supporting surrounding tissues. They are self-maintained and capable of repairing injury that could result from any excessive cyclic pressure loading (70 beats per minute, more than 100,000 beats per day, and more than 36 million beats per year) (Thubrikar, 1999 and Schoen and Levy, 1999).

Valvular heart disease is an important health issues instigating substantial indisposition and death worldwide (Jones et al., 2009). In fact, it is documented that the incidence of valvular disease escalates with age, ranging from 0.7% in the 18-44 years old group to 13.3% in the 75 years and older group (Nkomo et al., 2006; Mol et al., 2009 and Geemen et al., 2012), and this percentage is expected to increase due to an older and vaster population in the near future. Dysfunctional heart valve could result from inherited heart pathology, causing stenosis or inefficiency of the valve or could lead to diseases such as calcification, regurgitation, degeneration and stenosis or endocarditis. Endocarditis is a disease connected with the inflammation of endocardium, due to bacteria or an infection with rheumatic fever, and has a significant adverse effect on the functionality and long-term life of the patient.

Aortic valve is the most affected by valve diseases as it carries the maximum load of the heart. In general, however, the treatment of a dysfunctional heart valve is either repair or replacement with artificial one.

Prosthetic heart valves are instigated regularly to replace damaged natural ones, and they are extensively instigated in ventricular assist devices (VAD) and in total artificial hearts (TAH). These valves are of two kinds, mechanical with components fabricated from artificial materials such as carbon, metal, and polymers or tissue valves, which are assembled from either animal or human tissue, at least in part Literature suggests that over 285,000 substitute valves are employed per year, and roughly 40% of these are bio-prosthetic ones (Lin et al., 2004; Morsi et al., 2004; Mikos et al., 2006). Subsequently, due to a recent increase in the aging population, there has been a shift toward the increased use of tissue valves.

However, in general, all artificial valves suffer from a number of drawbacks that are different for mechanical and tissue ones. Nevertheless, the combination of valve-related complications is similar for both types of valves. The principal disadvantages of mechanical valves are mainly related to the significant risk of systemic thromboembolism and thrombosis which are attributed to the irregular blood flow through the rigid occluder, which aggravated by the use of artificial surfaces. Moreover there is always the risk of serious complications from hemorrhage associated with the necessary use of chronic anti-coagulation therapy (Schoen and Levy, 1999).

Oppositely, tissue valves maintain a low rate of thromboembolism exclusive of anti-coagulation therapy. However, these bio-prostheses, which are partly biologic tissue and partly synthetic are typically subject to progressive tissue deterioration which can lead to total structural dysfunction. Nevertheless, tissue valves demonstrate, in general, a beneficial potential of acclimatization to the patient's cardiovascular environment (Berry et al., 2010). Nonetheless, the main issues that hinder the progress of tissue valve substitutes include variations induced by preservation, construction, and methods of implantations. For example, it is known that the structural properties of tissues used in tissue valves are altered by glutaraldehyde cross-linking. Moreover, the choice of fixation technique is vital, as using the wrong technique could adversely affect the microstructur of the valve, particularly leaflets, and subsequently the degree of compliance compared to the biological tissue (Lin et al., 2004; Morsi and Birchall, 2005; Brody and Pandit, 2007). Generally, the degree to which the tissues are influenced contingent to the fixation technique used and the technique of construction. Moreover, the major cause of bio-prosthetic valve dysfunction is deterioration of cuspal tissue, which is primarily due to the inability of artificial materials to mimic the natural tissue. Calcification contributes to failure of tissue valves leading to regurgitation through tears in calcified cusps. This deterioration originates from mineralisation, often deep within the tissue (intrinsic) or at the surface (extrinsic) that are influenced by the host, the design of the implant and/or by the induced mechanical stresses. Non-calcified damage to the valvular structure that could accrue through constrained abnormal valve motion is also a major mechanism of degradation in porcine and pericardial prosthetic valves. Clearly, the preservation of tissue structural reliability is critical to the efficient performance and robustness of tissue heart valves (Schoen and Levy, 1999).

Conversely, the main drawbacks of artificial heart valves are the physical barriers and the restricted lifespan associated with them. The types and configurations of current artificial heart valves are well documented in literature, and a number of alternative designs have been suggested and examined structurally and hydrodynamically. Although the results from these studies revealed satisfactory correlation between *in vitro* measurements and *in vivo* clinical and pathological findings, they serve to highlight the continuing need for the development of long-term replacement valves (Morsi et al., 2007). In general, it is highly desirable that valve designs have durable membranes or leaflets and support structures that resist degradation due to *in vivo* environmental factors and mechanical stressing while retaining the functional characteristics and hemodynamics associated with the natural human heart valves. More importantly, to date the current valvular substitutes do not offer growth potential, which is a significant drawback for younger patients.

Recently, it has been recognised that an alternative to the fabrication of prosthetic valves is tissue engineering (TE) of an anatomically appropriate scaffold containing cells and controlling the development of normal functional valve architecture *in vivo*. This is indeed an exciting concept.

Figure 1.1 shows the concept of tissue engineering heart valves (TEHVs). This notion focuses on the creation of duplicate functional heart valve to eliminate all the current limitations of the clinical implementation of mechanical and bio-prosthetic heart-valves. However, existing results are still preliminary, and various issues remain to be addressed before the clinical application of a tissue-engineered heart valve will be possible for replacement (Deman, 2003, Morsi et al., 2005, Mendelson and Schoen, 2006; Mol et al., 2009). The three most important parts of tissue engineering are cells, biomaterials and conditioning techniques such as *in vitro* and *in vivo* (Migneco et al., 2008). Out of these three components, cells are the only living part, and hence the other two parts need to be optimized to suite the cells used for tissue generation.

Overall, TE of heart valves can be divided in two general strategies. The first strategy uses decellular (*not made up of or divided into cells*) *natural biomatrices*. With this decellularization (*remove cells or cellular material from an organ or tissue*) approach, the Extra Cellular Matrix (ECM) of say porcine heart valve is used as a scaffold (after elimination of their cellular antigens to reduce their immunogenicity) to guide the repopulation of the new cells after implementation in the patient. With this repopulation technique, various methods for decellularization have been proposed to ensure that the mechanical characteristics of the biomatrices or the alteration of the tissue *in vivo* is not distressed. In general, as will be discussed in Chapter 7, various concerns regarding the stability and durability of the natural biomatrices have to be addressed. Though, it has been debated that in comparison to biodegradable polymer scaffold TEHVs approach decellular biomatrices preserves natural ligands and Extra Cellular Matrix (ECM) constituents that could be more suitable for cell adhesion and proliferation (O'Brien et al., 1999, Brody and Pandit, 2007).

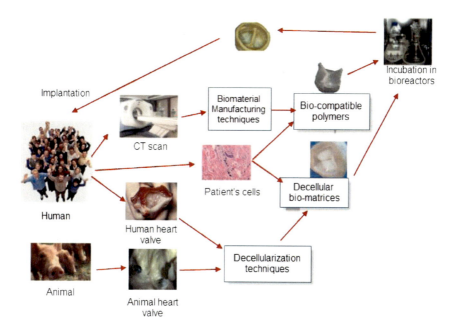

Figure 1.1. Illustration of the research steps of tissue engineering heart valve.

The second tactic is to use biocompatible *degradable nature or polymeric scaffolds* of the same geometry as valve rather than using decellularized valve. Generally, cells harvest are cultured and then seeded on these scaffolds, and after specific period for maturity *in vitro*, the resultant constructs are implanted *in vivo*. During this process the tissue develops into Extra Cellular Matrix (ECM) and the scaffold gradually degrades, leaving no trace of the synthetic materials. Ideally, eliminating immunological responses to the TE-construct and enabling the growth and remodelling to take place autologous cells should be used. However, other types of cells such as stem cells have been proposed. With this approach, the surface morphology of the scaffold is also critical to ensure good generation of the ECM, which in turn determines the overall structure and mechanical properties of the newly formed tissue (Shinoka et al., 1995, Hoerstrup et al., 2000, Bas, 2003). Moreover, with this approach, biological materials have been also used as scaffolds for TEHV alone or with combination of various types of polymers.

Still, *degradable polymeric scaffolds* approach became very popular recently and currently is the most favourable one by most researchers. Morsi et al. (Morsi et al., 2004; Morsi and Birchall, 2005) accentuated the urgent need to develop new materials for heart valves scaffolds, particularly leaflets that are capable of mimicking the deformation and coaptation of natural heart valves and called for hybrid approach, i.e., using artificial and natural materials, which will be discussed in Chapter 7. Moreover, further development is required in the areas of scaffold design, fabrication technology and biomaterials with suitable mechanical properties. However, while improvements are progressing, the manufacturing processes available to date still need to be optimised so that the scaffolds produced are suitable for tissue engineering of aortic heart valve. The use of harsh organic solvents for dissolving polymers and other chemicals required by the conventional fabrication process may produce toxic by-products. This in turn influences the biocompatibility of the valve constructs and elevates the issues of patient health (Morsi et al., 2008). Rapid prototyping technology, on the other hand, may offer an attractive and cost-effective approach. Nevertheless, it can be expected that continual research into the fabrication process will lead to advancements and breakthroughs in addressing the complex requirements of manufacturing scaffolds for tissue engineering of heart valve. Moreover, the developed scaffold must continually withstand the physiological conditions that natural valves are subjected to (engineering simulation and *in vitro* conditioning play important roles in this). Other issues that need to be addressed include understanding the factors that control cell adhesion, differentiation and proliferation, the optimal period of culture and its environment (static or dynamic) and methods of cell seeding (selective cell or mixed cell population seeding), which are vital for the development of a complex, tri-layered, flexible structure of the aortic valve that can function as a viable and efficient one. All these issues are fully discussed in this book in a systematic way.

1.2. The Book Structure

This book is organized into eight chapters; introduction and discussion on the research questions related to heart valve problems and the issues related to development of tissue engineering aortic heart valve are given in Chapter 1. Chapter 2, which provides the background information along with literature review on the clinical aspects of heart valve diseases treatments and diagnostics, follows this. Chapter 3 presents a full discussion of the types of heart valves replacements and their limitations. Chapter 4 discusses scaffolds materials natural and synthetics, and Chapter 5 presents an overview of the manufacturing techniques and gives a full discussion on the scaffolds manufacturing methods used for the creation of heart valve scaffolds. Chapter 6 discusses the *in vitro* conditioning and gives a review of the bioreactors and summarized the results of the Extra Cellular Matrix (ECM), followed by Chapter 7, which summarizes the research work related to tissue engineering aortic heart valves to date. Chapter 9 gives a summary of the issues of concern and recommendations for future research to create a functional tissue-engineered aortic heart valve.

References

Bas S. D., (Supervisor Y S Morsi), Design, development and optimisation of a tissue culture vessel system for tissue engineering applications, *Swinburne University of Technology*, Master Thesis, 2003

Brody, S. and A. Pandit, Approaches to heart valve tissue engineering scaffold design, *J. Biomed. Mater Res. B Appl. Biomater.*, 2007. 83(1): pp. 16-43.

Berry J. L., J. A. Steen, J. K. Williams, J. E. Jordan, A. Atala and J. J. Yoo, Bioreactors for development of tissue engineered heart valves, *Annals of Biomedical Engineering*, 2010. 38(11): pp. 3272-3279.

Geemen D. V., Driessen-Mol, A. Grootzwagers, L.G.M, Soekhradj-Soechit, R Riem Vis W., Baaijens, F. T., and Bouten C.V.C Variation in tissue outcome of ovine and human engineered heart valve constructs: relevance for tissue engineering *Regenerative Medicine*, 2012, Vol. 7, No. 1 , 59-7.

Jones, D. L., R. Adams, M. Carnethon, G. De Simone, T. B. Ferguson, K. Flegal, E. Ford, K. Furie, A. Go, K. Greenlund, N. Haase, S. Hailpern, M. Ho, V. Howard, B. Kissela, S. Kittner, D. Lackland, L. Lisabeth, A. Marelli, M. McDermott, J. Meigs, D. Mozaffarian, G. Nichol, C. O'Donnell, V. Roger, W. Rosamond, R. Sacco, P. Sorlie, R. Stafford, J. Steinberger, T. Thom, S. Wasserthiel-Smoller, N. Wong, J. Wylie-Rosett and Y. Hong, Writing group members,Heart Disease and Stroke Statistics—2009 Update: A Report From the American Heart Association Statistics Committee and Stroke Statistics Subcommittee, *Circulation,* 2009. 119(3): pp 480-486.

Hoerstrup, S. P., R. Sodian, S. Daebritz, J. Wang, E. A. Bacha, D. P. Martin, A. M. Moran, K. J. Guleserian, J. S. Sperling, S. Kaushal, J. P. Vacanti, F. J. Schoen and J. E. Mayer, Jr., Functional Living Trileaflet Heart Valves Grown In vitro, *Circulation,* 2000.102(90003): pp III-44-49.

Lin, Q., Y. S. Morsi, B. Smith and W. Yang, Numerical simulation and structure verification of Jellyfish heart valve, *International Journal of Computer Applications in Technology,* 2004.21(1-2): pp 2-7.

Mendelson, K. and F. J. Schoen, Heart valve tissue engineering: concepts, approaches, progress, and challenges, *Ann. Biomed. Eng.* 2006.34(12): pp 1799-1819.

Migneco, F., S.J. Hollister, and R.K. Birla, Tissue-engineered heart valve prostheses: 'state of the heart'. *Regen Med*, 2008. 3(3): p. 399-419.

Mikos, A. G., S. W. Herring, P. Ochareon, J. Elisseeff, H. H. Lu, R. Kandel, F. J. Schoen, M. Toner, D. Mooney, A. Atala, M. E. V. Dyke, D. Kaplan and G. V. Novakovic, Engineering Complex Tissues, *Tissue Engineering,* 2006.12(12): pp 3307-3339.

Mol, A., A. I. P. M. Smith, C. V. C. Bouten and F. P. T. Baaijens, Tissue engineering of heart valves: Advances and current challenges, *Expert Revof. Med. Devices.* 2009.6(3): pp 259-275.

Morsi, Y. S. and I. Birchall, Tissue engineering a functional aortic heart valve: an appraisal, *Future Cardiol.* 2005.1(3): pp 405-411.

Morsi, Y. S., I. E. Birchall and F. L. Rosenfeldt, Artificial aortic valves: an overview, *Int. J. Artif. Organs.* 2004. 27(6): pp 445-451.

Morsi Y. S., W. W. Yang, C. S. Wong and S. Das, Transient fluid-structure coupling for simulation of a trileaflet heart valve using weak coupling, *J. Artif. Organs.* 2007. 10(2): pp. 96-103.

Morsi, Y.S., Wong, C.S. and Patel, S.S., Conventional manufacturing process for three-dimensional scaffolds, *Virtual Prototyping of Biomanufacturing in Medical Applications,* 2008, Chapter7: pp. 129-148.

Nkomo, V. T., J. M. Gardin, T. N. Skelton, J. S. Gottdiener, C. G. Scott and M. E. Sarano, Burden of valvular heart diseases: a population-based study, *Lancet.* 2006.368(9540): pp 1005-1011.

O'Brien, M. F., S. Goldstein, S. Walsh, K. S. Black, R. Elkins and D. Clarke, The SynerGraft valve: a new acellular (nonglutaraldehyde-fixed) tissue heart valve for autologous recellularization first experimental studies before clinical implantation, *Semin. Thorac. Cardiovasc. Surg.* 1999.11(4 Suppl 1): pp 194-200.

Shinoka, T., C. K. Breuer, R. E. Tanel, G. Zund, T. Miura, P. X. Ma, R. Langer, J. P. Vacanti and J. E. Mayer, Jr., Tissue engineering heart valves: valve leaflet replacement study in a lamb model, *Ann. Thorac. Surg.* 1995.60(6 Suppl): pp S513-516.

Schoen, F. and R. Levy, Tissue heart valves: Current Challenges and future research perspectives, *Founders Aw 25th Annual Meeting Soc. Biomaterials.* 1999. Pp: 441-65.

Thubrikar, M. J., The aortic valve, Boca Raton, Florida, CRC Press, 1990.

Chapter II

Basic Introduction to Heart Valve Diseases

2.1. An Overview

AHA statistical update shows that in 2007, coronary heart disease caused about one of every six deaths in the United States with deaths of 406,351. Moreover, it is stated that each year, an estimated 785,000 Americans will have a new coronary attack, and approximately 470,000 will have a recurrent attack. Furthermore, it is projected that an extra 195,000 silent first myocardial infarctions would transpire every year, and every 25 seconds, an American will experience a coronary attack, and every minute approximately somebody will die of one of these diseases (Roger et al., 2011). Successively, each year, 795,000 persons suffer a new or chronic stroke. Furthermore, it cited that "from 1997 to 2007, the stroke death rate fell 44.8%, and the actual number of stroke deaths declined to 14.7%". In 2007, however, one in nine death certificates (277,193 deaths) in the United States were related to heart failure, and "the total number of in-patient cardiovascular operations and procedures increased by 27%, from 5,382,000 in 1997, to 6,846,000 in 2007"(Lloyed Jones et al., 2010; Roger et al., 2011).

Heart valve disease occurs when the valve becomes stenosed, restricting the flow of the blood or has a significant backflow (regurgitation) during diastole and is considered as one of the significant contributors to cardiovascular disease (CVD). Such disease can affect any of the four valves in the heart; however, it is most common in the aortic heart valve.

In developed, industrialised nations the most regular instigation of valve diseases has moved from rheumatic type to a degenerative pathology, affecting largely the middle-aged and senior people of age 40 years and over. Occurrence shows growing trend with cumulative age, with one in eight person aged 75 and older suffers from valve-related diseases (Mol et al., 2009). Consequently, due to the continuous growth of ageing population worldwide, the valvular heart diseases will no doubt substantially increase in the coming years. In developing countries, however, heart valve diseases mainly affect paediatric and younger patients, primarily due to the persistent burden of rheumatic fever. Though rheumatic fever seldom occurs, it is a serious disease and has a mortality of 2–5% and is responsible for many cases of damaged heart valves. The changing in the cardiovascular physiology alone or in combination with the effects of pathology, for example coronary artery disease, causes a reduction in elasticity and an increase in stiffness of the arterial of the heart with subsequent adverse effect on the functionality of the heart (Cheitlin 1991; Pugh and Wei 2001; Cheitlin 2003(A); Nichols 2005; Mol et al., 2009).

Moreover, some of main reasons of the cardiac valve dysfunction and calcification of the valve and its cusps are endocarditis, rheumatic fever, myxomatous[1] degeneration of congenital malformations, which often lead to stenosis and dysfunctional valve (Stewart et al., 1997; Morsi and Birchall 2005; Gelson et al., 2007; Carmo et al., 2010). Again, although one or several of these conditions could affect any of the four valves, due to significant hemodynamic conditions of the systemic circulation, the left-sided valves are particularly vulnerable, and the aortic valve is the most affected one (Brody, and Pandit; 2007 ; Sacks et al., 2009). Generally, depending on the degree of deterioration and complexity of the valve problem, the common treatment of the dysfunctional valve is repair rather than replacement. Hitherto, in the last few decades, heart valve replacements to overcome the problem of valve dysfunctional have gained considerable momentum.

In this chapter, we are mainly concerned with aortic valve disease, where the most common treatment for aortic valvular disease is replacement with prosthetic ones. Generally, there are different techniques, and various types of valves are implemented for replacement of a damaged heart valve depending upon the health and patient. However, although the techniques and skills for valve replacement have improved considerably, to date, there is no ideal valve replacement. An artificial valve that upholds regular valve mechanical and hemodynamics flow properties, non-thrombogenic and non-calcified has not yet been developed (Morsi and Birchall 2005; Brody and Pandit, 2007). This chapter starts with a brief description of the heart and valves anatomy, followed by discussion on the type of aortic heart valve diseases (AHVD), and summarizes valvular diseases and current diagnostic methods used.

2.2. Anatomy of the Heart and Valves

2.2.1. Human Heart

The heart is fabricated from muscle tissues, and its volume and weight function of age, sex, the size and the general health of the person. The grown-up male heart weighs approximately 325 grams, whereas the grown-up female heart weighs around 275 grams. Day by day, a normal heart contracts and beats about 100,000 times, while pumping nearly 7570 liters of blood. The heart is placed at the centre of the chest and consists of two boundaries, the left and the right and one is further divided into two chambers: the upper and the lower chamber (Figure 2.1). While the upper cavity, the atrium accumulates the blood, the lower cavity the ventricle pumps the blood out of the heart. The blood flow is maintained in a forward direction by the four valves through a synchronized action of opening and closing. The blood travels from the veins to the heart through the right atrium. This blood has large amount of carbon dioxide; however, it has relatively low oxygen, since it is absorbed by the

---

[1] Myxomatous degeneration as defined in Wikipedia refers to a pathological weakening of connective tissue. The term is most often used in the context of mitral valve prolapse, which is known more technically as "primary form of myxomatous degeneration of the mitral valve."

body tissues. When the heart filled with blood, the right atrium contracts, forcing the blood through the tricuspid valve, which opens into the right ventricle, and the blood is then pumped via the pulmonary valve and into the lungs. The blood, which is now enriched with oxygen and depleted of carbon dioxide, returns to the left atrium. Subsequently, it is pumped through the mitral valve into the left ventricle and finally, the blood is pumped out through the aortic valve into the aorta and the rest of the circulatory system (Bender, 1992; Bas, 2003; Consilient-health 2012).

The main purpose of the four valves is to regulator the blood flow through the heart and, in general, the human heart has two kinds of valves, namely bicuspid and tricuspid, reflecting the numbers of leaflets within the valve. The valves in the heart are unidirectional to stop backflow (regurgitation) of blood from one chamber to another and have sufficient flexibility and mechanical integrity to stand all the hemodynamics forces and pressure fluctuations during each heart cycle.

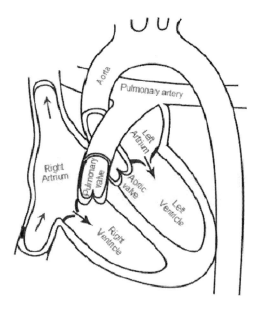

Figure 2.1. Schematic illustration and cross section of a human heart with blood flow directions.

2.2.2. Function, Anatomy and Locations of Heart Valves

Within the heart, there is one pair of atrioventricular valves and semilunar valves. These are briefly discussed below:

2.2.3. Atrioventricular Valves (AV Valves)

The atrioventricular valves (AV) are positioned amid the atria and ventricles; the one on the right is recognized as the tricuspid valve and the one on the left is known as the mitral valve. The atrioventricular valves have leaflets, chordae tendineae, and papillary muscles.

The leaflets are thin fibrous membranes, which bind the area of the opening between the atria and ventricles. They are white, thin and luminous, and intensely robust, with vigorous tissues, and when these leaflets are distressed they become thick, dense and occasionally calcium becomes accumulated on them. Moreover, the papillary muscles are attached to the valve leaflets through the chordae tendineae, which are the stem of heart muscle that emerge from the wall of the ventricle, and these muscles unremittingly contract and are responsible for preserving the valve leaflets sealed during the time of ventricle contraction. When the pressure within the ventricles is elevated, the valve seals and averts the blood from escaping out to the atria (Mancini 2011).

The left AV valve has two papillary muscles, whereas the right valve has three. Note again that the chordae tendineae are thin cords that bind the papillary muscles to the valve leaflets, and they are non-elastic and perform an significant function in transferring the forces of the papillary muscle to the valve leaflets. In general, each leaflet has about 5 to 60 chordae attached to it, and these chordae are inserted into the leaflet from the papillary muscle giving sufficient opening for blood to flow from the atria to ventricles. A brief description of function of left and right Atrioventricular valves (AV) are given below (Mancini 2011):

A. Tricuspid Valve

This valve is a right-sided AV valve and is positioned between the right atrium and ventricle to inhibit the backflow of blood as it is propelled from the right atrium to the right ventricle. It consists of three leaflets, one interior, one inferior and one septal, which are attached to the inter-ventricular septum. Due to the right side location of tricuspid valve, it is not exposed to significant hemodynamic forces and as such has low risk of valvular diseases and only endocarditis disease can affect it with fatal consequences.

B. Mitral Valve

This valve is left-sided AV, and its function is to prevent the backflow of blood coming from the left atrium to the left ventricle and guard the opening between the left atrium and ventricle. The mitral valve consists of two leaflets, posterior and anterior leaflets, where the former is longer and narrower than the anterior one. Due to the fact that sufficient high pressures and hemodynamic forces are generated on the left side of the heart, the mitral valve is the most single stressed one, after the aortic one, and as such, diseases in mitral valve are very common.

2.2.4. Semilunar Valves (SV)

The semilunar valves (SV) are two valve arrangements that rest among the right ventricle and the pulmonary artery and between the left ventricle and the aorta. The valves leaflets open and close to regulate the flow of blood in one direction only out of the heart, while avoiding backflow of blood. The aortic valve is set between the left ventricle and aorta and the pulmonary one is located in the middle of the pulmonary artery and right ventricle. Both valves are flap-like structures and unidirectional types that allow blood to flow in one direction. The classification of the semilunar group valves is discussed in the following sections (Mancini 2011)

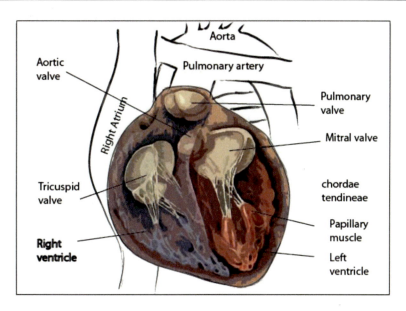

Figure 2.2. Schematic artistic illustration of the four heart valves within the heart created by the "Illustrator."

A. Aortic Valve

The function of this valve is to prevent the backflow of blood as it is pumped, allowing a proper amount of blood from the left ventricle to the aorta, and it permits the progress of oxygenated blood from the left ventricle into the aorta giving the needed oxygen to the body and its tissues. Dysfunctional aortic valve allow very little blood passing through and force the blood to remain in the ventricle, which may result in left ventricular bulge and deteriorating the ventricular muscles with adverse consequences.

B. Pulmonary Valve

This valve located between the top of the right ventricle and the pulmonary artery, and its function is to prevent the backflow of blood as it is pumped from the right ventricle to the pulmonary artery. The moment the right ventricle contracts during systole phase the valve allows the blood to flow from the ventricle into the pulmonary artery, and subsequently the blood continues its way from the artery into the lungs where it is oxygenated. However, in the middle of every heartbeat, the leaflets remain shut so that blood from the pulmonary artery cannot flow back to the right ventricle i.e. preventing regurgitation phenomena to occur. Figure 2.3 shows the cross section of the heart and the location of the four valves.

2.2.5. Tissue Structure, Extra Cellular Matrix (ECM) of Aortic Valve

In general, the tissue structures, Extra Cellular Matrix (ECM) and the overall function of the all four valves within the heart are very analogous. However, as stated previously, the aortic valve is the most durable; during systole, the leaflets rest upon the aortic wall and close efficiently and passively during diastole and throughout the cardiac cycles. Still, the leaflets of the valve go through a significant number of deformations and coaptation cycles, and they do possess complex and cyclical structural rearrangements and specialized, functionally adapted Extra Cellular Matrix (ECM) (Mendelson and Schoen, 2006).

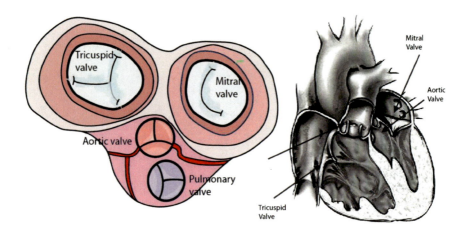

Figure 2.3. Schematic illustrations of the locations of the four valves and connecting arteries, using the "Illustrator."

The aortic valve is a semilunar heart valve, which has three layers: (1) ventricularis (2) spongiosa and (3) fibrosa. These layers have the properties that make the heart valve adjust for its function. Interstitial cells colonize the matrix of heart valves and express a variety of phenotypes, and a large proportion of these cells communicate smooth muscle cells. Moreover, interstitial cells are sufficiently thin to be perfused from the heart's blood, and they maintain the ECM. In addition, out of the three diverse layers, the fibrosa keeps the main muscle, whereas the spongiosa facilitates the relative interchange among the two fibrous layers and observe the bending and deformation energy during closure. What is more, fibrosa is mainly composed of circumferentially aligned collagen fibres, spongiosa is abundant in proteoglycans and loosely arranged collagen and the ventricularis consists of radially arranged aligned elastin fibres . Moreover, the function of the elastin of the ventricularis is to aid the leaflets to reduce the surface area when the valve is fully open and expand these leaflets when blood pressure is applied (Bas 2003; Mol et al., 2005).

It should be noted that the *aortic valve leaflets* are lined on both the aortic side as well as the ventricular side by the endothelial cells, which perform as a protective layer of the valve leaflets. These leaflets are primarily avascular, and in case of aortic valve, the pressure differential across the closed valve imposes a large hemodynamic load on the fibrous network within the leaflets, which transport these forces to the aortic annulus (Bas 2003; Mendelson and Schoen 2006; Misfeld and Sievers 2007). Furthermore, note that the contraction and the deformation of the leaflets during the heart cycle assist in sustaining the hemodynamic functionality of opening and closing of the valve and providing support during the diastole phase (Dagum et al., 1999).

Moreover, the cells present in the three layers of the aortic valve leaflets are known as the interstitial cells. There are three types of interstitial cells present there: (A) smooth muscle cells, (B) fibroblasts and (C) myofibroblasts. The smooth muscle cells are normally present separately or in bundles and are distributed in the fibrosa of the valve leaflets. The fibroblasts are the ones that mainly responsible for maintaining the Extra Cellular Matrix (ECM) of the valve leaflets and they are mainly present in the ventricularis of the leaflets. The myofibroblasts, on the other hand, are the cells with phenotypic features of smooth muscle cells as well as the fibroblasts and are also present along with the smooth muscle cells in the fibrosa. These distributions of the interstitial cells change based on their biological and mechanical microenvironments (Mendelson and Schoen 2006; Misfeld and Sievers 2007; Mancini 2011).

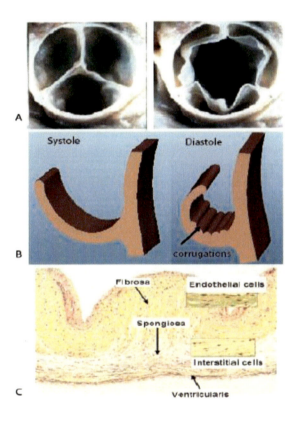

Figure 2.4. (A) Photograph of the aortic valve in open and closed position (from the aorta) (B) Biomechanical cooperatively between elastin and collagen during valve motion (C) Aortic valve histology emphasizing tri-laminar structure and presence of valvular interstitial and endothelial cells (Mendelson and Schoen, 2006, with permission).

2.3. The Cardiac Cycle

The cardiac cycle epitomises the successive events that arise during a single cycle of heartbeat. It is principally divided into two major phases: Diastole and Systole ones. Diastole characterizes the range of time when the ventricles are relaxed and throughout this period, blood is inactively flowing from the left and right atriums to the left and right ventricles, respectively. The blood floods via mitral and tricuspid valves that divide the atria from the ventricles. The right atrium acquires venous blood from the body through the superior vena cava and inferior vena cava, and the left atrium receives oxygenated blood from lungs through four pulmonary veins that enter the left atrium. At the end of diastole phase, both atria contract; this accelerates an extra amount of blood into the ventricles (Kalbunde, 2011).

Systole phase signifies the time during which the left and right ventricles contract and eject the blood into the aorta and pulmonary artery respectively. Throughout the systole phase, the aortic and pulmonary valves release to allow the discharge of blood into the aorta and pulmonary artery. The atrioventricular valves are shut during systole phase, so no blood is flowing to the ventricles, but continues to go into the atria through the vena cava and pulmonary veins. The cardiac cycle is divided typically into seven phases which are Atrial Contraction, Isovolumetric contraction, Rapid Ejection, Reduced Ejection, Isovolumetric Relaxation, Rapid Filling and Reduced filling, and these are fully discussed in literature (see, for example, Kalbunde, 2011). Here, only a brief account is reproduced to the main five stages of each beat of the cardiac cycles illustrated in Figure 2.5. These stages are:

- **Late diastole**—the semilunar valves close ; the Aortic valve opens, and the whole heart eases.
- **Atrial systole**—Atria is contracting; AV valves open, and blood floods from atrium to the ventricle.
- **Isovaleric ventricular contraction**—where the ventricles commence to contract; AV valves shut, as well as the semilunar valves with no change in volume.
- **Ventricular ejection**—where the ventricles are empty; they are still contracting, and the semilunar valves are open.
- **Isovaleric ventricular relaxation**—when the blood pressure decreases with no blood is going through the ventricles. Ventricles end contracting and commence to relax; semilunars are locked as the blood flow in the aorta is forcing them sealed.

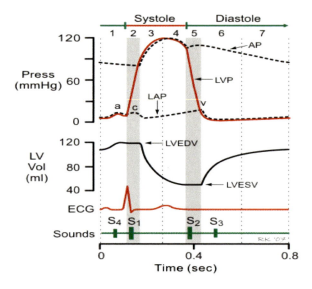

Figure 2.5. Schematic illustration of cardiac cycle (Kalbunde, 2011).

Moreover, it should be noted that during the complete heart cycle, the blood pressure rises and declines, which is due to the effect of specialised electrical pulse provided by heart cells sino-atrial node and the atrioventricular node. The cardiac muscle is synchronized with myocytes, which instigate their own contraction without any support of the external nerves. It should be noted that cardiac cycle of a healthy heart takes approximately one second.

2.4. Valvular Heart Disease (VHD)

2.4.1. Overview

Heart valve disease or valvular heart disease is a condition where heart valves do not execute effectively their blood regulation task. This can occur in any of the four valves, Tricuspid, Pulmonary, Mitral and Aortic valves or any grouping of them. As stated above, heart valves in general have tissue leaflets that open and close as a function of the cardiac cycle to control blood flow within the heart. A dysfunctional valve transpires when the valve fails to open or close effectively due to regurgitation, stenosis and atresia disease. There are numerous publications related to VHD, which were well discussed and reviewed by Bender, 1992; here, however a brief account is reintroduced.

The major aggravation of heart valve disease is congestive heart failure; it is a condition that occurs when the heart is not able to pump out an acceptable volume of blood. When blood accumulates in the veins, it causes congestion of fluid in the body tissues, which affects lungs and other parts of the body. The accumulation of fluid in the lungs leads to the obstruction of flow of air and oxygen exchange and interferences with breathing. In addition, the upsurge of fluid in the legs and swelling of the ankles occur, which leads to breathlessness and eventually causes congestive heart failure. Some other symptoms include palpitations, fatigue, fainting, chest pain and heart muscle diseases. Moreover, when the blood clots and these clots separate, they block the blood stream, which is called emboli. This is a serious issue since these little pieces of freed clots usually get stuck in a small blood vessel, thus preventing the affected organ from receiving sufficient blood to function probably. This phenomenon is particularly significant in the brain as a clot in it can lead to a stroke. On the other hand, clots in the legs can cause severe pain or in the extreme cases, gangrene (Bender, 1992).

2.4.2. Infective Endocarditis (Bacterial Endocarditis)

Infective endocarditis is an inflammatory condition, which can lead to heart valve disease. It is an infection of the endocardium, which is the lining that covers the inner walls of the heart's chambers and the valves (Consilient-health, 2012). Bender 1992 reported that this infection occurs when bacteria, fungi, or microorganisms continue to develop on the valves' inner surface and produce small, warty nodules. Two types of bacteria, streptococci and staphylococci, are often responsible for causing this condition, and it is generally referred to as bacterial endocarditis. Common symptoms of this disease are fatigue, mild fever, weakness and joint-aches. Microorganisms that grow on the endocardium generally cause holes in the valves, may distort its shape and even disrupt its function entirely. The groups of infectious microorganisms could initiate the development of emboli. These emboli subsequently flow with the blood circulation and stop the flow of the blood in vary small arteries and vessels.

The infection can progress to critical levels and may even cause death due to heart failure (Bender, 1992).

2.4.3. Myxomatus Degeneration

Myxomatus degeneration affects mainly the elderly. This process, which normally affects the mitral valve, results from a series of metabolic changes that cause the valve's tissue to lose its elasticity, become weak and flabby, and become covered up by an accumulation of starch deposits (Bender, 1992).

2.4.4. Calcific Degeneration

Calcific generation is a process wherein calcium deposits accumulate on the valve, and this tissue generation invariably causes aortic stenosis, which narrows the aortic valve. It can also affect the mitral valve, resulting in it becoming leaky or regurgitating (Bender, 1992).

2.4.5. Congenital Anomalies

Congenital abnormalities are problems that are present from birth, and these can lead to heart valve diseases. Malformed aortic valve is the most common congenital defect in which the aortic valve has only two leaflets instead of the normal three, and hence is called bicuspid. Although this defect can be rectified through surgery, such action is attempted only if there are symptoms of, or repeated, valve infections (Bender, 1992).

Voss et al. (2007) reported that the frequency at which the acquired heart valve diseases occur is as follows (Voss et al., 2007).

- Aortic Valve, 65%
- Mitral Valve, 30%
- Pulmonary and tricuspid Valves disease, 5%

As stated above, in general diseases related to the tricuspid and pulmonary valves are rare and the most frequent and critical valve complications occur in the mitral and aortic valves respectively. In this book, we are mainly concerned with aortic valve, and its diseases are briefly discussed below.

2.5. Diseases of Aortic Valve

2.5.1. Overview

As stated previously, the function of the aortic valve is to regulate the flow of the blood from the left ventricle through the outflow tract of the aortic vessel through to the ascending aorta to supply the oxygenated blood to the rest of the body. This valve consists of a complex structure and has three leaflets, which are semilunar in shape. The semilunar structure of the valve makes it ideal for dealing with the large pressure fluctuation exerted upon the valve during systole. As fully discussed by Mihaljevic et al. (2008), each leaflet of the aortic valve is a dilatation of the ascending aortic wall, termed the aortic sinus, and the ostia of the coronary arteries instigate from two of the coronary sinuses, with the aortic vestibule being placed just below the valve. Aortic valve is slightly below the pulmonary valve, which located to its left. It has a similar structure to the pulmonary valve but has a stronger assembly to keep standing the high blood flow and pressure on the left side of the heart, and its composition is superior to the mitral valve. As shown in the Figure 2.6, the most common features of the aortic valve include the three valve leaflets, annulus, part of the fibrous skeleton of the heart and the sinuses of valsalva (Mihaljevic et al., 2008).

It should also be noted that the aortic valve is passive device that can be opened and closed with minimal pressure difference between the ventricle and aorta. In the closing period, the fully integrated cusps assembly is aligning perfectly in order to resist the backflow pressure into ventricle. Hence, the line of the valve closure is always below the free edge of the leaflets (Mihaljevic et al., 2008). This type of valve is known to suffer from two major diseases, namely *Aortic Regurgitation* and *Aortic Stenosis*, and these are summarized here.

2.5.2. Aortic Regurgitation (AR)

This type of dysfunctional influences aortic root and leaflets and adversely affects the closure of the valve and is primarily due to annular enlargement and/or occasionally due to valve prolapse. It is very common in between the ages of 30 and 60 and is less common during early childhood but can also develop over years. In the case of chronic aortic regurgitation, there occurs an increase in stroke volume, which subsequently initiates hypertension, high pulse pressure and increased after-load. This increased after-load could be similar in magnitude to that occurring in aortic stenosis (Shipton and Wabha, 2001).

Literature findings also suggest that aortic regurgitation is a degenerative disease as small amount of calcifications could be observed but rarely with condensed myxomatous tissue *(a change in the structure or chemical composition of a tissue or organ that interferes with its normal functioning)*. Aortic regurgitation can also be generated by bicuspid aortic valve leading malcoaptation and occasionally prolapse of the largest leaflet (Enriquez-Sarano et al., 2008). Other causes of aortic regurgitation, though uncommon, may involve Marfan syndrome, a connective tissue heritage disease where patients have elongated weak bones and extremely malleable joints. Moreover, it should be noted that rheumatic aortic regurgitation commences retraction of the valve leaflet, leaving a central regurgitate orifice, and is difficult to repair. With this, morphologic assessment should be utilized when clearly defining the mechanism of aortic regurgitation.

Moreover, it is well known that aortic regurgitation leads to overload of volume and pressure of the left ventricle, which adjusts by both concentric and eccentric variation till a point of after load disparity, when left ventricular (LV) systolic function corrupted. In this case an assessment of LV dimension and systolic function by echocardiography is indispensable. Any level of LV dysfunction justifies rescue surgery but can result in decreased post-operative survival (Cohn, 2008; Enriquez-Sarano et al., 2008). Figure 2.7 shows schematic illustration of aortic valve regurgitation.

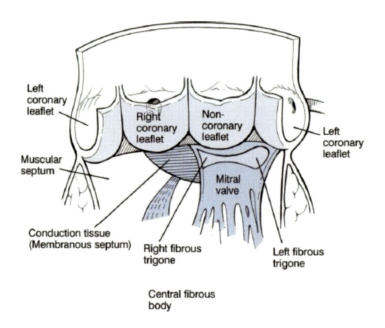

Figure 2.6. Schematic of aortic valve and sinus valsalva at aortic junction (Mihaljevic et al., 2008, with permission).

2.5.3. Aortic Stenosis (AS)

Aortic stenosis is defined as a narrowing or blockage of the aortic valve opening, and it is usually idiopathic and results in calcification and degeneration of the aortic leaflets. It is inexorable that people born with bicuspid aortic valve are prone to develop aortic stenosis at some point in their lives. Although there is a variation in the rate of progression of aortic stenosis from patient to patient, still nearly 75% of them die within three years after the onset of symptoms, if replacement of the aortic valve with artificial one is not implemented. It is observed, however, that some patients with severe aortic stenosis remain asymptomatic, while patients with even moderate stenosis have symptoms that characterize this condition. It should be noted that the average area of normal aortic valve is 2.5 cm^2 and with no pressure gradient. Hence, if the valve area is found to be < 0.8 cm^2 and/or if the pressure gradient is > 50 mm Hg, then the degree of stenosis is considered to be critical with clear symptoms (Shipton and Wabha, 2001).

Moreover, it should be noted as illustrated in Figure 2.8, the with aortic stenosis, the valve leaflets become coated with deposits which change the structure of the leaflets with an adverse effect on valve performance. Recently it has been recognized that there is a strong link between aortic stenosis (AS) and atherosclerosis which emphasis the significant of cholesterol deposition.

Basic Introduction to Heart Valve Diseases 19

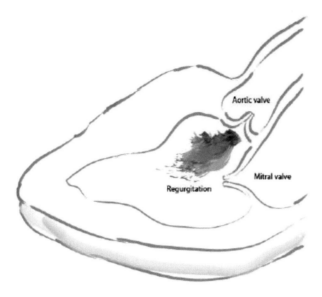

Figure 2.7. Schematic of Aortic Regurgitation.

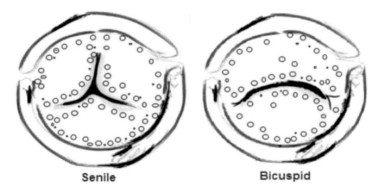

Figure 2.8. Schematic illustration of Senile and Bicuspid Aortic Stenosis.

2.5.4. Bicuspid Aortic Valve

This is a heart condition that is due to a congenital deformity and is one of the most common congenital heart defects, affecting about 1-2% of all children. As noted above, a normal aortic valve has three leaflets to control the flow of blood through the heart, whereas bicuspid valve has only two leaflets and, as such, the flow of the blood through it is somewhat restricted. The negative impact of a bicuspid aortic valve can significantly vary from patient to patient. There can be acute aortic stenosis, which can be either inherited or generated. The significant of the symptom of bicuspid aortic valve gradually increases with age and reaches its peak value around the age of forty. Moreover, with severe aortic stenosis, the hypertrophy of the left ventricular arises as well as decreasing wall stress and sustaining the left ventricular ejection fraction (Cheitlin, 2003(B)). Still, in general, for unidentified factors bicuspid valves are more likely to calcify than tricuspid ones and advanced AS are difficult to clinically identify and assess (Enriquez-Sarano et al., 2008).

2.6. Valvular Heart Disease and Diagnostic Techniques

As stated above, some of the most significant types of heart valve problems include displacement of the valve and narrowing of the valve due to mainly valve *stenosis or regurgitation* of the valve. Moreover, these types of heart valve problems can be due to several reasons, acquired from rheumatic fever or could be developed over the years. When the problem becomes severe and starts to interfere with the blood flow, the patient often suffers from the problem of irregular heartbeats, frequent problem of fainting and breathing, etc.

For the evaluation and monitoring of the valvular disease, the noninvasive imaging techniques have been widely used. The methods have almost completely replaced angiocardiography (using x-rays following the injection of a radiopaque substance) for the diagnosis and determination of the seriousness of valvular heart disease (Cheitlin, 1991; Bas, 2003). However, it should be stated that the initial step of diagnostic in the determination of a VHD remains cardiac auscultation. Moreover, the literature suggests that the ultimate and popular technique currently implemented is echocardiography which can diagnose various types of heart valve-related diseases (Cheitlin, 2003(A); Bas, 2003). Among the oldest non-invasive techniques available are the modified X-ray procedures, which can be used to analyze softer tissue, including blood vessels lungs and or intestines. Although current X-ray techniques are not precisely used for the detection of valvular disease, they are able to identify various types of abnormalities in the blood vessels, heart to facilities the diagnostics of valvular disease.

Current restraints of MRI are first visualization of implanted metallic valves is not possible due to the magnetic field generated by this technique and it is very expensive to operate and use. Generally, CT and MRI evaluation of patients with VHD provide vital morphologic and physiologic evidence related to status of the disease. In fact, examination of heart chambers and aortic artery dimension as well as ventricular wall thickness stipulate the foundation for identifying the seriousness of the disease being examined (Rozenshtein and Boxt, 2000; Bas, 2003). Still, for evaluation of the degree of stenosis, determination of trans-valvular pressure gradient is a suitable means, and MRI might not reflect any advantages over echocardiography (Garbi, 2011; Barasch, 2011; Docstoc, 2011). However, CT and MRI diagnostics techniques are basically never used in the initial analysis of HVD.

Furthermore, both ultra-fast CT (UFCT) and MRI create high-resolution cardiac images, but (UFCT) needs intravenous injection of x-ray contrast medium, while MRI does not (MacMillan, 1992; Bas, 2003). Purposely, MRI is very practical for determination, evaluation, and semi-estimation of valvular regurgitation, whereas ultra-fast CT is not (Docstoc, 2011). Another major disadvantage with CT is that radiation can harm foetal tissue. However, although both diagnostic methods can evaluate aortic and mitral valve stenosis and determine coronary artery bypass graft conditions in general, ultra-fast CT is the chosen technique (MacMillan, 1992; Bas, 2003; Papanikolaou et al., 2010). Table 2.1 summaries diagnostic techniques used for HVD and their main features and applications.

Table 2.1. Non-invasive techniques for the assessment of VHD

Diagnostic Techniques	Investigators	Features	Applications
Echocardiography	Cheitlin, 2003 (A), Garbi, 2011, Barasch, 2011	Ultrasound waves are used to produce images	Heart murmur, enlargement and infections can be detected
X-ray	Mini Physics*	Excitement and de-excitement of electrons in the selected atoms produces X-ray.	In addition to the detection of irregularities in heart and blood vessels, it can also be used in imaging of soft tissues, e.g., lungs, blood vessels and intestines.
CT Scan	Health Guide**	Imaging technique where X-rays are used to produce cross-sectional images of the human heart.	It gives excellent images and can be used to detect aortic and mitral valve stenosis and also offers various features for the imaging and sizing of the heart.
MRI Scan	Papanikolaou et al., 2010	Unrestricted viewing angle and has excellent degree of precision.	Generates high resolution and cardiac images of excellent quality and offers various features for the sizing of heart. Good in the detection of valvular regurgitation and aortic and mitral valve stenosis as well as CAB graft. .

*http://miniphysics.blogspot.com/2010/11/features-of-x-ray-spectrum_27.html
**http://health.nytimes.com/health/guides/test/ct-scan/overview.html
CAB: Coronary Artery Bypass

References

Bender, J. R., *Heart Valve Disease,* Chapter 13, Heart Book, Yale University of Medicine, 1992. pp. 167-176.
http://www.scribd.com/doc/39785094/HEART-VALVE-DISEASE, [Viewed on 06.07. 2011].

Barasch E., W*hy Doctors use Echocardiography,* Heart, The Doctor Will See You Now, 2011, from: http://www.thedoctorwillseeyounow.com/content/heart/art2032.html, viewed on: 12/10/11.

Bas S. D., (Supervisor Y S Morsi), Design, development and optimisation of a tissue culture vessel system for tissue engineering applications, *Swinburne University of Technology,* Master Thesis, 2003

Brody, S. and A. Pandit, *Approaches to heart valve tissue engineering scaffold design, J. Biomed. Mater Res. B Appl. Biomater.,* 2007. 83(1): pp. 16-43.

Carmo, P., M. J. Andrade, C. Aguiar, R. Rodrigues, R. Gouveia and J. A. Silva, *Mitral annular disjunction in myxomatous mitral valve disease: a relevant abnormality recognizable by transthoracic echocardiography, Cardiovasc. Ultrasound.,* 2010.8(53).

Cheitlin, M. D., *Valvular heart disease: management and intervention. Clinical overview and discussion, Circulation,* 1991, 84 (3 Suppl): pp. 1259-1264.

Cheitlin, M. D., *Cardiovascular physiology-changes with aging, Am. J. Geriatr. Cardiol.,* 2003 (A), 12(1): pp.9-13.

Cheitlin, M. D., *Pathophysiology of valvular aortic stenosis in the elderly, Am. J. Geriatr. Cardiol.,* 2003 (B), 12(3): pp. 173-177.

Cohn, L. H., *Cardiac surgery in the adult*, McGraw-Hill, New York. 2008.

Consilient-health, viewed 2012, http://www.consilient-health.com

Docstoc, 2011, *Heart Disease and Artificial Heart Valves*, Chapter 2.
http://www.docstoc.com/docs/32815302/CHAPTER-2-HEART-DISEASE-AND-ARTIFICIAL-HEART-VALVES-21 [viewed on: 02/12/2011].

Dagum, P., G. R. Green, F. J. Nistal, G. T. Daughters, and T. A. Timek, "*Deformational Dynamics of the Aortic Root : Modes and Physiologic Determinants." Circulation,* 1999, 100 (90002): pp. 1154-1162.

Enriquez-Sarano M., Nkomo V. T., Michelena H., Principles and Practice of Echocardiography in Cardiac Surgery, Cohn Lh, ed, Cardiac Surgery in the Adult. New York: McGraw-Hill, 2008:315-348. Chapter 11.

Garbi M., *The general principle of echocardiography*, The EAE Textbook of Echocardiography, Chapter 1 2011.
http://escecho.oxfordmedicine.com/cgi/content/abstract/1/1/med-9780199599639-chapter -001, [Viewed on: 12/10/11].

Gelson, E., M. Gatzoulis and M. Johnson, *Valvular heart disease,* Brit. Med. J., 2007. 335(7628): pp. 1042-45.

Health Guide, *CT Scan,* The New York Times, 2011, from:
http://health.nytimes.com/health/guides/test/ct-scan/overview.html, [Viewed on: 12/10 /11].

Klabunde R. E., *Cardiovascular Physiology Concept.* 2011, http://www.cvphysiology.com /Heart%20Disease/HD002.htm,

[Viewed on: 12/10/11].

Mancini M. C., *Heart Anatomy,* Medscape reference. 2011, http://emedicine.medscape.com/article/905502-overview#aw2aab6b6, [Viewed on: 12/10 /11].

Mini Physics, *Features of X-ray spectrum,*2011, from: http://miniphysics.blogspot.com/2010/11/features-of-x-ray-spectrum_27.html, [Viewed on: 12/10/11].

MacMillan R. M., *Magnetic resonance imaging vs. ultrafast computed tomography for cardiac diagnosis, Int. J. Card. Imaging.* 1992, 8(3): pp. 217-227.

Mendelson, K. and F. J. Schoen. *Heart valve tissue engineering: concepts, approaches, progress, and challenges, Ann. Biomed. Eng.,* 2006, 34(12): pp. 1799-1819.

Mihaljevic, T. I., S. Paul, L. H. Cohn and A. Wechsler, *Pathophysiology of Aortic Valve Disease,* Cardiac surgery in the Adult, McGraw-Hill Professional, 2008. pp: 825-840.

Misfeld, M. and H. H. Sievers., Heart valve macro- and microstructure, Philos. *Trans. R. Soc. Lond. B: Biol. Sci.,* 2007, 362(1484): pp. 1421-1436.

Mol, A. D., M. I. van Lieshout, G. C. Dam, S. P. Hoerstrup, F. P. T. Baiijensand C. V. C. Bouten. Fibrin as a cell carrier in cardiovascular tissue engineering applications, *Biomaterials,* 2005, 26(16): pp. 3113-3121.

Mol, A., A. I. P. M. Smits, C. V. C. Bouten and F. P. T. Baaijjens, *Tissue engineering of heart valves: advances and current challenges, Expert Rev. Med. Devices,* 2009. 6(3): pp. 259-275.

Morsi, Y. S. and I. Birchall, Tissue engineering a functional aortic heart valve: an appraisal, *Future Cardiol.,* 2005. 1(3): pp. 405-411.

Nichols, W. W., Clinical measurement of arterial stiffness obtained from noninvasive pressure waveforms, *Am. J. Hypertens.,* 2005 18(1 Pt 2): 3S-10S.

Pugh, K. G. and J. Y. Wei., Clinical implications of physiological changes in the aging heart, *Drugs. Aging,* 2001, 18(4): pp. 263-276.

Papanikolaou V., M. H. Khan and I. J. Keogh, Incidental Findings on MRI Scans of Patients Presenting with Audiovestibular Symptoms, BMC Ear, *Nose and Throat Disorder*, 2010. 10(6).

Roger, V. L., A. S. Go, D. M. L. Jones, R. J. Adams, J. D. Berry, T. M. Brown, M. R. Carnethon, S. Dai, G. D. Simone, E. S. Ford, C. S. Fox, H. F. Fullerton, C. Gillespie, K. J. Greenlund, S. M. Hailpern, J. A. Heit, P. M. Ho, V. J. Howard, B. M. Kissela, S. J. Kittner, D.T. Lackland, J. H. Lichtman, L. D. Lisabeth, D. M. Makuc, G. M. Marcus, A. Marelli, D. B. Matchar, M. M. MacDermott, J. B. Meigs, C. S. Moy, D. Mozaffarian, M. E. Mussolino, G. Nichol, N. P. Paynter, W. D. Rosamond, P. D. Sorlie, R. S. Stafford, T. N. Turan, M. B. Turner, N. D. Wong, J. W. Rosett, Heart Disease and Stroke Statistics_2011 Update: A Report From the American Heart Association, *Circulation,* 2011. 123: pp. e18-e209.

Rozenshtein, A. and L.M. Boxt, Computed Tomography and Magnetic Resonance Imaging of Patients With Valvular Heart Disease. *Journal of Thoracic Imaging,* 2000. 15(4): p. 252-264.

Sacks, M.S., F. J. Schoen and J. E. Mayer, Bioengineering challenges for heart valve tissue engineering, *Annu. Rev. Biomed. Eng.,* 2009. 11: pp. 289-313.

Stewart, B. F., D. Siscovick, B. K. Lind, J. M. Gardin, J. S. Gottdiener, V. E. Smith, D. W. Kitzman and C. M. Otto, Clinical factors associated with calcific aortic valve disease. Cardiovascular Health Study, *J. Am. Coll. Cardiol.,* 1997. 29(3): pp. 630-4.

Shipton, B., and H. Wabha, Valvular Heart Disease: Review and Update, Am. Fam. Physician. 2001 Jun 1; 63(11): pp. 2201-2209. http://www.aafp.org/afp/2001 /0601 /p2201.html [Viewed on 06.07.2011].

Voss, A., R. Schroeder, A. Seeck, and T. Huebner., Biomechanical Systems Technology, Volume 2: Analysing Cardiac Biomechanics by Heart *Sound.* Ed: C. T. Leondes, World Scientific Publications. 2007

Chapter III

Heart Valve Replacements

3.1. Overview

As pointed out in previous chapters, valvular heart diseases (HVD) signify a major public-health problem causing significant sickness and death worldwide. However, paediatric and young-adult patients are mostly affected by HVD in the developing countries, whereas in industrialized nations, the diseases affect mostly the elderly persons. Still, even though various surgical techniques were developed to mend dysfunctional heart valves,70% of the diseased ones are difficult to repair and the most frequent and effective treatment is valve replacement. Moreover, it is stated that approximately 290,000 heart valve substitutes are implemented annually worldwide, and the number of patients needing valve replacements is projected to almost triple in the near future (Mol et al., 2009). There is no doubt that artificial heart valve replacements have saved the lives of so many patients, however, it is never considered to be a perfect alternative for the replacement of dysfunctional heart valve. On the other hand, the world population will increase to around 8.9 billion in 2050, and the need for heart valve replacement annually around 850,000 by 2050. In addition, the largest population increase will be seen in third world countries, and due to shortage of resources and cost, most of the patients who need heart valve replacement will face a lot of difficulties (Yacoub and Takkenberg, 2005; Mol et al., 2009).

As previously stated, human heart valves are reflexive devices that open and close in reaction to changes in heart pressure to sustain the unidirectional flow of blood through the heart. They are extremely efficient when they function, probably with minimal or no resistance to advance blood flow and allowing just small amount of backflow when closed. With dysfunctional heart valve, an artificial heart valve replacement offers an enhanced cardiovascular function, long-term endurance and quality of life.

Moreover, it is very clear from literature that design and developments of artificial heart valve substitutes have gone through a lot of development. A snapshot of the development of these devices shows that in 1950s, the American surgeon Dr. Hufnagel was the first to implement artificial heart valve for the treatment of malfunctioning one. Then, later in 1960, with the support of cardiopulmonary bypass, the first implant of artificial valve in the anatomic location took place, and since then, the field of cardiac valve design has developed considerably by the opening of various pioneering surgical and cardiac additive techniques as well as optimised model of mechanical and bio-prosthetics valves that have helped in saving hundreds of thousands of lives worldwide (Gudbjartsson et al., 2008). However, in spite of many developments over the years, in valve design and several hundred different configurations of valve substitutes that have been considered, there is still, to date, no perfect one. The majority of the proposed designs have been neglected due to problems discovered during *in vitro* and *in vivo* evaluations. The major contributing factor is the fact that the non-homogeneity of the natural valve structure makes it enormously challenging to develop homogenous prosthetic material that is capable of satisfying all the mechanical and biological characteristics for a well-organized function of the valve. As such, no an artificial material that is capable of even moderately mimicking the functional performance and the durability of the natural tissues has yet been developed. Moreover, when compared to nature valve in terms of regularity of the blood flow without regurgitation, haemolysis, or excessive stresses in the leaflets and sustaining tissues structure, current valve substitutes have a range of limitations and disadvantages. Subsequently, ideal valve replacement has yet to be developed, and the current general practice is a conciliation to meet the needs of an individual patient. Still, a vital setback inbuilt in the use of existing artificial valves in the paediatric patients is their limitation to grow, repair, and integrate with the host (Sacks et al., 2009). However, as stated in Chapter 1, it is trusted that engineering autologous tissue heart valve would effectively address all these particularly those related to remodeling and growth, as will be discussed in Chapter 7 of this book (Sacks et al., 2009).

This chapter gives a brief summary of some of the main characteristics of artificial heart valves, mechanical and tissue ones that are currently used and highlights their major clinical limitations. Particular focus is given to aortic heart valve and its current performance from structure and hemodynamic reliability point of view.

3.2. Type of Valve Replacements

There are two groups of heart valve substitutes available in the market, namely mechanical and biological/tissue valves, and generally these are of two principal types:

- Mechanical Prosthetics, which include:

  - Caged Ball valves
  - Tilting Disc Valves
  - Bi-leaflet Disc Valves

- Tissue/ Biological Valves, which include:

  I. Xenografts, Hetrografts
 II. Homograft
III. Autografts with Ross Procedure

Each one of the above-listed valves has certain characteristics and limitations, which are snappishly summarized below:

3.2.1. Mechanical Valves

These valves are made of arrangement of metal and elastomeric parts and designed to fulfill the functionality of the natural valves. There are various designs and configurations of prosthetic mechanical valves available in the market. Generally, the material chosen to assemble mechanical prosthetic heart valves are alumina, titanium, carbon, polyester and polyurethane. In selecting the materials, various important factors are considered including mechanical integrity, biocompatibility and ease of manufacturing. Moreover, compression and tensile forces, twisting, shear **stresses, Poisson's ratio, resistance** and flexibility are some of the mechanical properties considered in selecting materials. Therefore, valve materials need to perfectly adjust to those frequently changing forces within the heart. Modulus of elasticity of the selected material determines its favorability by the elastic deformation in some areas, and yield stress point is important for the material selected, which can make the difference between elastic deformation and failure.

As listed above, mechanical valves are of three main configurations: *Caged-ball, tilting-disk valve and bi-leaflet valve*. In an effort to deal with some of the drawbacks of these valves, they have undergone various developments and modifications of valve configurations and materials, which are briefly highlighted below.

3.2.1.1. Mechanical Caged-Ball

Figure 3.1 shows schematic of *caged-ball valve*, which is constructed from a curved metal cage supported by a suture ring that houses a silicone elastomeric ball (Pick, 2010). This type of valve works on the basis that elastomeric ball dislodges due to pressure difference from ventricle and aorta during the contraction and relaxation of the heart. Dr. Hufnagel, an American surgeon who was the first to implement an artificial heart valve in a human, invented the caged-ball valve. Subsequently, in 1960, with the aid of cardiopulmonary bypass, the first implant of artificial valve in the anatomic position took place. Later in 1960, surgeons Lowell Edwards and Albert Starr developed another version of the caged-ball, which was implemented in an orthotopic position.

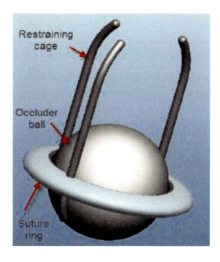

Figure 3.1. Pro/Engineer 3D illustration of Mechanical Caged-ball.

Although the caged-ball valve suffers from a poor hemodynamics, it had initially experienced a lot of success, and it was the standard valve substitute for over four decades (Shiono et al., 2005; Suezawa et al., 2008). Subsequently, various modifications to the design of this valve were proposed, including the change of materials of the ball from Silastic type to Stellite, geometric change on the cage curves, suppressing of the ball occlude and adding extra material coating to the sewing ring and the cage, as well as modification in the sewing ring itself (Emery et al., 2008).

However, despite all these modifications to the caged-ball valve, the occurrence of thrombus formation and blood cells damages owing to high shear stress, stagnation, and flow separation were still common (Rajani et al., 2007, Tarzia et al., 2007, Dasi et al., 2009). Moreover, outflow obstructions of this type of valve have been observed around the orifice of the swing ring, the gap between the cage and the walls of ascending aorta, and around the ball, which in most cases showed abnormal flow distributions. With such poor performance, the recipient of this type of valve had to use a significant amount of anticoagulation (Ezekowitz, 2002, Grados et al., 2007).

3.2.1.2. Tilting-Disk Valve

The tilting disc valve was mainly designed to improve up on the poor hemodynamic performance of caged-ball type of valves which were bulky and heavy to implement. Subsequently, in an attempt to improve the hemodynamic including the pressure drops across such valve, a flat disc, instead of a ball, was developed as a caged-disc valve. The first tilting disk valve was proposed in 1969, known as Bjork-Shiley valve. This tilting disk design had a single circular occluder, which was made of an extremely hard carbon material that was supported by two metal struts. With this design, the opening angle of tilting disk was 60°, and it closed shut completely at a rate of 70 cycles/ minute. Subsequently, this type of valve had gone through a number of materials and geometrical modifications until 1980, as illustrated in Figure 3.2.

However, in general, it is well accepted that Bjork-Shiley valve offers a much better hemodynamic performance and better functionality than a caged-ball valve and creates less damage to blood cells with minimum blood clotting. Nevertheless, Bjork-Shiley still suffers from imperfect hemodynamic and a degree of flow circulations at the minor flow orifice, which could lead to embolization of the disk. Furthermore, these types of valves are known to suffer from outlet struts fractured due to the cyclic stress that causes colliding of the struts by the disk.

Subsequently, some of these problems have been partially overcome by the introduction of the second generation of tilting-disc known as the Medtronic Hall valve (B). Figure 3-2 shows the Pro/Engineer drawings of Bjork-Shiley valve (A) and the modified Medtronic Hall valve (B), respectively.

The third generation of disc valve, St. Jude Medical (SJM), was invented and proposed by Nicoloff and associates in 1979 (Emery et al., 1979; Gott et al., 2003). As shown in Figure 3.3, it consists of two semicircular leaflets that revolve about the struts that are attached to the valve housing. It should be noted, however, that in general, St. Jude Medical valve offers a number of advantages over other proceeding disc type valves. These include better hemodynamics, a greater effective opening area (2.4-3.2 cm^2), and low aortic pressure drop, and they can be implemented with less anticoagulation consumption with minimum thromboembolism rate (Hanssen et al., 2003; Wang et al., 2003; Mohty et al., 2006). Later, others modifications to SJM valves have been carried out and as a result, the third generation bio-leaflet aortic valve was introduced as Sulzer Carbo Medics valve, the ATS Medical prosthesis, and the On-X prosthesis (Gudbjartsson et al., 2003).

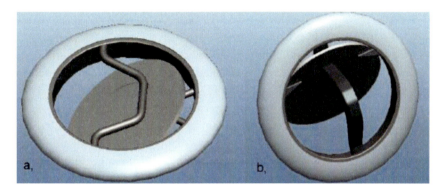

Figure 3.2. Pro/Engineer illustration of Bjork-Shiley valve (a) and the modified Medtronic Hall valve (b).

In general, SJM types of valves, due to their superior performance and functionality (Butchart et al., 2001; Khan et al., 2001), have been widely implemented for many decades (Kortke and Korfer, 2001; Butchart et al., 2002). In 1993, Carbo-Medics introduced an enhanced version of the St. Jude model (Figure 3.3b). This design had two pyrolite discs, which are semicircular in shape and open at an angle of 78°, i.e., 18° more than Bjork valve. Moreover, a titanium-stiffening ring is also incorporated in this design that ensures the prevention of the two discs being dislodged. Additionally, these valves have an effective and large orifice area with quite efficient hemodynamic behaviour in terms of hemodynamics and low degree of turbulence generated in comparison with others similar mechanical valves.

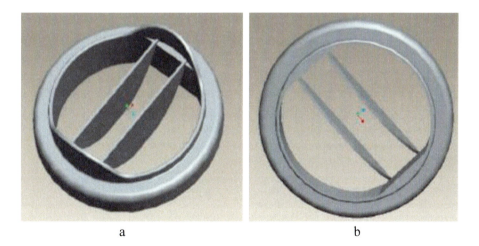

Figure 3.3. A Pro/Engineer Illustration of SJM types of valves, St Jude Medical (A) and Carbo Medics Bi-leaflet Valves (B).

A timeline progression of mechanical heart valves prosthesis is shown in Figure 3.4.

3.2.2. Tissue Bio-Prosthetic Valves

As outlined in Section 3.2, these types of valves are classified into Homograft (also called allograft) and Autograft and Ross operation and Xenografts, which have different characteristics and configurations as briefly examined below.

3.2.2.1. Homografts or Allograft Valves

Homografts or allografts are human tissue valves, donated by a deceased person, that are dissected and removed from the donor, then treated with antibiotics before being transplanted to the most suitable and compatible patient. In general, homograft heart valve transplant patients do not have to undertake anticoagulant therapy, and these valves exhibit a more reliable hemodynamic performance than the other types of valves (Clinic, 2011). Moreover, generally speaking, with these valves, there are no major drawbacks regarding rejection by the recipient's body, and they do not necessitate any type of immune suppressive treatment. In addition, this type of valve has excellent hemodynamic performance, low incidence of thromboembolic complications and does not require chronic anticoagulation, and, as such, they are highly suitable options for valve replacement. Still, they offer a low risk of infection and are less thrombogenic and they have superior biological and chemical compatibility and, therefore, longevity. Nevertheless, in general, there is still a essential matter of concern regarding all the cryogenically preserved tissue valves as they do not autonomously grow, and at the time of transplant, they cannot be implemented in the optimum desirable positions. Moreover, availability of such type of valves is restricted due to the scarcity of the donors, which makes them a very expensive option for most of the patients worldwide, and they can at times be hostile and cause immune response in children (Clinic, 2011).

With respect to the historical development of these valves, it is documented that the first attempt of using homograft was made in 1956, by Gordon Murray, a heart surgeon who treated an aortic insufficiency by implanting the homograft with the heterotopic position in the descending aorta (Murray, 1956; Murray, 1960). The results were satisfactory for the following four years of implementation, which has been repeated and confirmed by other surgeons (Mohri et al., 1966; Nelson et al., 1967). However, the first attempt for implanting the homograft valve in orthotopic position failed and was subsequently corrected and successfully implanted shortly afterward by Donald Ross, Barratt-Boyes and Paneth and O'Brien at the same time (Ross, 1962; Boyes, 1964; Paneth and O'Brien, 1966; Shumacker, 1992). From this point onward, the homograft prosthesis implantation became a very successful practice as an aortic heart valve replacement, with a substantial amount of research being carried out on the improvement and development of preservation techniques (Sands et al., 1967, Boyes, 1971, and O'Brien et al., 1987).

3.2.2.2. Autografts and Ross Procedure

Autografts, are the process of the healthy functioning valves being used to replace the malfunctioning one within the body of the patient. Transplantation of a living valve from the pulmonary position to the aortic position within the same person was proposed, as the two valves are structurally similar. This pulmonary autograft operation was first described as the *Ross procedure*, and previous studies showed that such valves remain viable and functional in their new position without degenerating for approximately 13 years. The important feature of this approach is that the valve used is living, autogenous and of optimal structure.

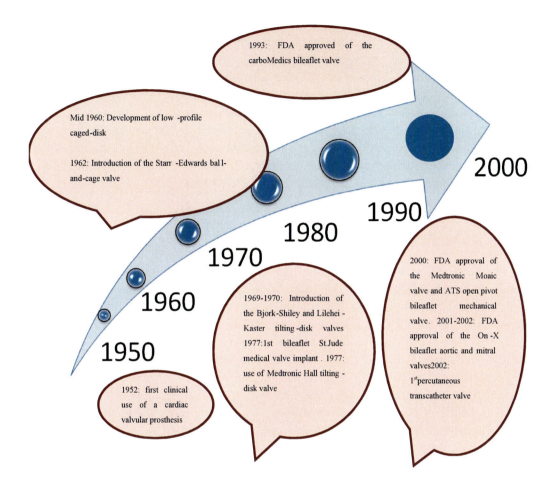

Figure 3.4. Timeline progression of mechanical heart valves prosthesis.

The Ross procedure technique is named after its inventor, Donald Ross, and it is the most widely accepted procedure for autografts. The technique comprises of replacing a malfunctioning native aortic valve with the host's **own pulmonary valve** and then using another homograft or artificial one to replace the pulmonary one. The rationale behind this technique is that using an identical valve from the same patient guarantees no pressure drops and offers efficient hemodynamic and reliable performance. More importantly, the valve would remodel, which is particularly important with children, as the valve will continue to grow with the recipient child. Moreover, other biological tissues, of both human and animal **origin, have been used to replace malfunctioning aortic valves with Ross's operation.** The most noteworthy of these include human fascia lata and bovine pericardium. The fascia lata **valve was constructed at the time of operation using the patient's own living tissue** and was first developed in 1966, with encouraging results (Michell et al., 1998).

However, it should be pointed out that with autograft and Ross operation, the procedure is highly complex in nature and requires highly experienced surgeons to carry it out. In addition, the compatibility of the dimension and size of a pulmonary valve at the new location of an aortic valve is always a matter **of main concern.** Although Ross's procedure is **very** convenient, it is risky, very complicated and there are some controversial results about the **uncertainty over whether the grafts will grow commensurate with the recipient's growth.**

Moreover, the leakage from the valve is a very common problem with these kinds of valves and may require a subsequent operation. Other complications are aortic regurgitation and stenosis of the coronary artery, and since artificial heart valve is transplanted into the pulmonary position, there is always the chance right-sided endocarditis might occur. Though, it should be noted that a pulmonary tissue valve replacing an aortic valve could lead to failure of the valve due to the fact that the hemodynamic environment is quite different for pulmonary valves to endure in alternate positions. Moreover, it has been argued in literature that with Ross procedure, one cannot ascertain exactly the true advantage of this technique as a comparison among the performance delivered by a Ross transplant and other artificial tissue valves such as porcine or pericardial valves (David et al., 1996).

3.2.3. Xenografts or Hetrografts

The third unique classification of tissue valves is xenograft valves, also known as hetrografts, which originated from the heart tissues of an animal (e.g., pig or cow) and then sorted out and preserved using glutaraldehyde solution. Commonly used xenografts are known as porcine from pig's heart tissues and bovine from cow's heart tissues. Porcine valves are constructed by sewing the tissue on the cobalt-nickel alloy wire, and then the cloth skirt is sewed to the wire (Figure 3.5). The structure of the bovine pericardial valve is very similar to that of porcine valves, the only difference being the presence of a small metallic cylinder connecting the circumferential wire's ends. The metal wire arrangement in both porcine and bovine are said to occupy large amount of space, which could easily add up to the orifice area and provide more space for the blood flow. However, porcine tissue valves are known to have better longevity, and in general, bovine pericardial valves are often considered less favourable option over porcine xenografts (Hammermeister et al., 1993).

However, literature suggests that the implantation of valves constructed from bovine pericardium has been successful when moulding fresh tissue to a tricuspid configuration around a support frame formed such valves. While firmly held in this position, the tissue is fixed with glutaraldehyde (Black et al., 1983). The most popular valve of this type was the Ionescu-Shiley valve, which was introduced into clinical use in 1976, but was subsequently drawn from the market in 1987, due to some serious complications.

However, regardless of the tissue source, these valves can be configured in two different ways: stented and stentless. Figure 3.5 shows a photograph of used xenografted valve. Figure 3.6, on the other hand shows Pro/Engineer 3D drawing of the concept of (a) stented (b) stentless.

The historical development of tissue valves stated in 1969, when Kaiser and Hancock introduced a new form of valve replacement using an explanted, chemically treated, porcine aortic valve mounted on a cloth-covered metal or plastic frame. This type of valve became commercially available in 1971, and was introduced as Hancock porcine tissue valve. Other valves of this type have since been developed, and the most popular model was known as the Carpentier-Edwards, which was later withdrawn from the market in 1990 (Desai et al., 2009).

In the stented valves, leaflets are sewn into three U-shaped metal wire frames that are made from a cobalt-nickel alloy (Figure 3.6). All of these frames are attached to a basement called swing ring, which is covered by a Dacron cloth sewing skirt. The fixation of the leaflets on the stent can be performed at different pressures with the assistance of glutaraldehyde (a tissue fixative). Unlike pericardial valves which are fixed in low- or zero-pressure conditions, porcine ones can be fixed at different pressure readings, high (60 to 80 mm Hg), low (0.1 to 2 mm Hg), as well as zero-pressure (0 mm Hg) respectively (Desai and Christakis, 2008; Desai et al., 2009).

Figure 3.5. Photo of used xenografted Valve.

Later, the novel design of the aortic valve replacement called stentless valves came from the idea of replacing the entire aortic root and adjacent aorta as a block. In 1990, Tirone David proposed and developed the first stentless porcine valve at St. Jude Medical, Inc., St. Paul, MN (David et al., 1990), called Toronto SPV. The initial approach was to replace the extracted glutaraldehyde-preserved porcine septal muscles from the donor into the aortic position after trimming the sinuses of Valsalva in the recipient. The external polyester fabric from the annulus to the top of each commissure covered the new assembly. Although this new design eliminated some problems associated with the frame and the swing ring, it introduced unpredictable challenges for implantation techniques such as size of aortic root. Therefore, the initial technique was expanded to the replacement of the entire aortic root, which led to production of new stentless AVR, such as Aortic Root Bioprosthesis (Medtronic Inc., Minneapolis, MN), and the Edwards Prima (Edwards Life sciences LLC, Irvine, CA).

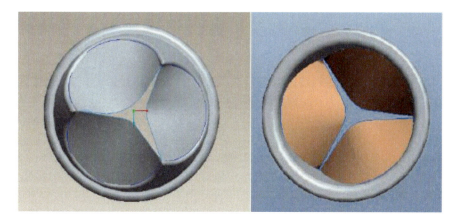

Figure 3.6. Pro/Engineer illustrations of (a) stented (b) stentless.

Subsequently, the first comparative clinical data was presented by Cohen based on postoperative hemodynamic results obtained by echocardiography in Carpentier-Edwards pericardial valves and Toronto stentless porcine valves used in two different groups of patients with the same aortic root. The results showed no difference in indexed effective orifice area (IEOA) or left ventricular mass (LVM) regression between the two groups and also showed the same functional outcomes within the first year (Cohen et al., 2002). Similar findings from the other researchers confirmed the negligible difference in hemodynamic results of stented and stentless porcine, which seriously challenged the benefit of stentless valve complex construction. However, Borger et al. (2005) showed modestly lower mean gradients in stentless prostheses versus stented prostheses (9 mm Hg versus 15 mm Hg) and LVM (100 g/m^2 versus 107 g/m^2) in their nonrandomized study.

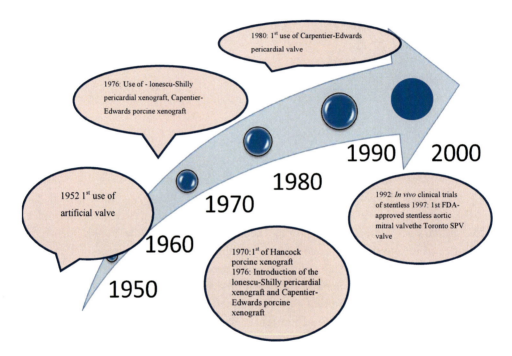

Figure 3.7. Summary of the timeline progression of biological heart valves prosthesis.

The timeline progression in the biological heart valves prosthesis has been summarized as shown in Figure 3.7.

3.3. Complications and Limitations of Artificial Heart Valves: An Overview

Currently, the replacement of diseased or damaged valves with artificial substitutes is becoming a routine clinical exercise, and the discipline of heart valve replacements has become one of the major growth areas in cardiac surgery. However, the artificial heart valves are not ideal, and, despite the huge demand for heart valve substitutes and the introduction of hundreds conceptual designs over the years, the majority of these designs have been discarded due to problems found during preclinical evaluations and trials (*in vitro* and *in vivo* experimentations). Subsequently, patients who undergo a heart valve surgery by present techniques require re-operation in 10-15 years in almost 50-60% of patients. Although tissue heart valve implantation does not require anticoagulant treatment, their clinical significant deteriorations are frequently encountered after four to five years of implantation. Furthermore, they are less durable than mechanical valves, and other problems such as leaflets mineralisation with or without secondary tearing and mechanical failure and fatigue of the cuspal structure were reported with these valves (Schoen et al., 1982; Schoen and Levy, 1999).

Mechanical valves, on the other hand, were shown to be more durable, but patients with mechanical valves still require anticoagulant therapy throughout their lives, as thrombus formation in the swing ring and behind the valve is a common problem in these types of valves, which may break off and embolise to various parts of the body. In addition, for women of childbearing age, mechanical valves are unsuitable as the anti-coagulant used can potentially cause birth defects in the first trimester of foetal development. Still, with mechanical valves, regurgitation occurs during the closed phase of the valve cycle as most these valves possess clearance between the disc edge and the annulus to stop possible jamming of the disc. Likewise, leakage occurring in the gaps induces high shear stresses, and if cumulated with the turbulent flow out of the valve, then these leakage surge would be nearly four times greater than the maximum forward flow velocity (Knott et al., 1988). Besides, one of the major concerns with mechanical heart valves is the primary elements, which cause damage to red blood cells (RBCs). These RBCs comprise of flexuous membranes and haemoglobins, and their ability to carry oxygen establishes the proper functioning of blood. Red blood cells can easily alter their shapes without tampering the membranes. When the cells shape is changed to the extreme, the RBCs get damaged due to shear stress and lose their flexural property. This phenomenon of RBC bearing the irreparable damage and haemoglobin loss due to membrane rupturing is called haemolysis, and it is considered to be a major drawback for mechanical valves (Sallam and Hwang, 1984). Moreover, in literature, there are many controlled experiments done to evaluate the duration and magnitude of shear stress, which can cause haemolysis. The literature suggests that shear stresses of the order of 1500 dynes/cm^2 to 4000 dynes/cm^2 cause irremediable damage to RBCs (Blackshear et al., 1965and Morsi et al., 1999). Moreover, if the time duration of the shear stress is lower, then the damage to RBCs can also be expected to be reduced. Hence, it is recommended that shear stress generated by an artificial valve should always be as low as 500 dynes/cm^2. It has also been pointed out that leakages occurring through the closed gaps are a bigger concern than the wearing off that occurs due to forward flow (Ellis et al., 1998).

However, it can be stated that in spite of many enhancements to the design specifications of various artificial valves to reduce the turbulent backflow, all of the progress has done little improvement to patient's health and the long-term suitability of mechanical heart valves are still subjected to questioning.

With tissue valves, all the evidence to date signifies the advantages of using stented biological prosthesis in the patients with small aortic root on better left ventricular mass (LVM), resulting in regression results and decreased thromboembolic events (Dellgren et al., 2002; Gelsomino et al., 2002). However, it might be argued that using the stentless biological prosthesis could be more beneficial for younger patients with small aortic roots. In essence, less elevated residual gradient would be created in comparison to stented biological prosthesis cases, which gives more freedom to the patient for physical activities. Moreover, the subsequent researches showed—regardless of the complexity of implantation—excellent haemodynamics and a significant decrease in the thickness of the heart after using stentless porcine aortic valve, regardless of the complexity of implantation. Still, it is very difficult to ascertain with certainty whether the stentless porcine valve is better than the stented porcine ones, and the research is ongoing (Dellgren et al., 2002).

Moreover, in general, the use of xenograft tissue valves poses a major threat of infecting the patients with diseases stemming from the animal's body such BSE and scrapie diseases, which are very fatal. Currently, there are no effective biological or chemical diagnostic techniques to detect the remains of these diseases in the tissue valve before implementation. In addition, healthy time span offered by a xenograft in comparison to mechanical valves is not considerably high. Others problems related to valve infection of endocarditis and non-structural dysfunction that could occur in the two types of artificial heart valve, tissue and mechanical ones (Schoen and Levy, 1999). However, it is expected that these problems will be reduced by further modifications of the material and/or by using new materials. Currently, there are numerous investigations aiming at developing leaflet valves from artificial materials such as polyhydroxyalkanoate biopolyester or modified polyurethanes (Sodian et al., 2000).

In general, however, with respect to the use of tissue valves, the major concerns are the problems of calcifications caused by calcium deposition and valvular failure due to inability of the valve structure to autonomously grow (Schoen et al., 1992). The degeneration of the tissue valves is a function of time and often occurs at a very consistent rate. Still, the consistent calcium deposition on the valve leaflet can cause the blood flow to stall, leading to the problem of stenosis or, to a complete damage of the tissue leaflets. As such, with high amounts of calcium intake and metabolism in growing patients, such as children, not all types of tissue valves are considered as a favourable option; hence, this option, tissue valves, remains viable for older patients only. Obviously, most of the complications associated with tissue valves are due to their inability to mimic the natural tissue performance during the heart cycles and that induces substantial and repetitive stresses, resulting in increasing tissue fatigues. Moreover, current fixation and sterilization such as glutaraldehyde affect structural integrity of the tissue. Hence, it is highly desirable that valve designs incorporate both durable membranes for leaflets and have support structures that resist degradation from environmental and mechanical stressing factors, while retaining the functionality and the mechanical integrity of the natural valve. One line of attack is to mainly address the fixing processes for bioprosthetic valves and develop new techniques that can provide an advancement of the resilience of the material and diminish the occurrence of dystrophic calcification and do not alter the comparatively nonthrombogenic nature of existing bioprostheses. Another line is to focus on the development of new long-lasting, nonthrombogenic materials, for example, alumina, titanium, polyetheruretane pyrolitic and so on (Gonzalez et al., 2003).

However, in a nutshell, despite all the enhancements and major strides that have taken place in the design field of artificial heart valves in general, a design that assures a complication–free result for the patients is still to be determined. Many monitoring and pre-diagnosis technologies are being proficiently manufactured worldwide with the aim of preventing thrombosis, haemolysis or stenosis from occurring with artificial heart valves.

Table 3.1 summarizes the limitations and advantages of artificial heart valves.

Table 3.1. Features of different types of valves
2007

Vales	Investigators	Characteristics and Configurations	Advantages	Limitations/comments
Caged-ball Valve	Ezekowitz, 2002, Shiono et al., 2005, Tarzia et al., 2007, Grados et al., 2007 Suezawa et al., 2008, Emery et al., 2008, Rajani et al., 2007, Dasi et al., 2009,	Curved metal cage, held by suture ring with a silicone elastomeric ball is located in the center of the valve. The pressure difference due to the transformed phase of ventricle and aorta ensure functionality. This valve experienced a number of modifications to materials of the ball and the configurations of valve.	This valve was the standard valve substitute and enjoyed success for almost 40 years.	The recipient must take high degree of anticoagulation. Unsatisfactory hemodynamic performance that including, thrombus formation, blood cell damage, high shear stress, stagnation and flow separation.
Bjork-Shiley valve	Emery et al., 1979, Gott et al., 2003	Made of hard carbon material, with a single occluder of an opening angle of 60°.	Offers better functionality than caged-ball valve with limited blood cell damage and clotting.	Complications such as eddy related flow formation, embolization of the leaflet of the valve and outlet struts fractured can occur.
St. Jude Medical Valve	Emery et al., 1979, Butchart et al., 2000, Khan et al., 2001, Kortke and Korfer, 2001, Butchart et al., 2002 Gott et al., 2003,	Consists of two semilunar leaflets, which are attached to the valve housing. It has good mechanical integrity and functionality.	Better hemodynamics including larger effective opening area, less anticoagulation and lower thromboembolism rate.	Strut fracture that might cause other complications can occur.
	Hanssen et al., 2003, Wang et al., 2003, Mohty et al., 2006,			
Carbo-Medics	Kortke and Korfer, 2001, Gudbjartsson et al., 2003,	Have two semilunar pyrolytic discs. An improved design of St. Jude valve. A titanium ring is added so that the discs cannot be dislodged.	Consists of effective and large orifice area. Opening angle is more than Bjork Shiley valve and efficient hemodynamic characteristics with low degree of turbulence.	Strut fracture can occur, and it might cause other complications as well.

Table 3.1. (Continued)

Vales	Investigators	Characteristics and Configurations	Advantages	Limitations/comments
Homograft/Allograft Valve	Sands et al., 1967, Boyes, 1971, O'Brian et al., 1987	This is a human tissue valve that is capable of providing biological and chemical compatibility and longevity.	This valve has a number of advantages including requires no anticoagulation therapy, has low risk of infection and gives reliable hemodynamic performance with lower thromboembolic complications.	Difficulties at the time of transplant, as they do not autonomously grow. Limited availability of these organs, and thus it is expensive.
Autograft Valve	Michell et al., 1998, David et al., 1996	It is living, autogenous and of optimal structure and Ross procedure is performed for these valves.	Excellent hemodynamic performance and reliability with no pressure drops.	The procedure is highly complex and requires highly experienced surgeons.
			Can remain viable and functional in their new position without degeneration and can remodel.	Compatibility of the dimension and the size of a pulmonary valve at the location of an aortic valve can present a challenge. Leakage from valve, aortic regurgitation, stenosis, right-sides endocarditis might occur.
Xenograft/ Hetrograft valve	Hammermeister et al., 1993, Schoen and Levy, 1999, Sodian et al., 2000	Originated from the heart tissue of the animal, and glutaraldehyde solution is used to preserve. Metal wire arrangements occupy large amount of space. The new valve used polymers to advance better reproducibility and low cost.	Orifice area increased, hence more space for the blood flow.	Bovine pericardial valves are less favourable than porcine xenografts. Valve infections might occur due to the use of animal tissues.

References

Boyes, B. G. B., Homograft Aortic Valve Replacement in Aortic Incompetence and Stenosis, *Thorax,* 1964. 19(2): pp. 131-150.

Boyes, B. G. B., Long-term follow-up of aortic valvar grafts, *British Heart Journal,* 1971. 33, Supplement: pp. 60-65.

Borger, M.A., S. M. Carson, J. Ivanov, V. Rao, H. E. Scully, C. M. Feindel and T. E. David, Stentless Aortic Valves are Hemodynamically Superior to Stented Valves During Mid-Term Follow-Up: A Large Retrospective Study, *Ann. Thorac. Surg.,* 2005. 80(6): pp. 2180-2185.

Butchart, E. G., H. H. Li, N. Payne, K. Buchan and G. L. Grunkemeir, Twenty years' experience with the Medtronic Hall valve, *J. Thorac. Cardiovasc. Surg.,* 2001. 121: pp. 1090-1100.

Butchart, E. G., N. Payne, H. H. Li, K. Buchan, K. Mandana and G. L. Grunkemeier, Better anticoagulation control improves survival after valve replacement, *J. Thorac. Cardiovasc. Surg.,* 2002. 123: pp. 715-723.

Blackshear, P. L. Jr., F. D. Dorman and J. H. Steinbach, Some mechanical effects that influence hemolysis, *Trans. Am. Soc. Artif. Organs,*1965. 11: pp. 112-117.

Black, M.M., Drury, P.J., and Tindale, W.B., Twenty-five years of heart valve substitutes: A review. *Journal of the Royal Society of Medicine,* 1983. 76(8): p. 667-680.

Cohen, G., G. T. Christakis, C. D. Joyner, C. D. Morgan, M. Tamariz, N. Hanayama, H. Mallidi, J. P. Szalai, M. Katic, V. Rao, S. E. Fremes and B. S. Goldman, Are stentless valves hemodynamically superior to stented valves? A prospective randomized trial, *Ann. Thorac. Surg.,* 2002. 73(3): pp. 767-775.

Clinic C., Aortic valve surgery in the young adult patient, 2011. Written with Gosta Pettersson, MD, Cardiovascular & Thoracic Surgery, Miller Family Heart & Vascular Institute from [http://my.clevelandclinic.org/heart/disorders/valve/youngvalve.aspx] viewed on 11/11 /11.

Dasi, L. P., H. A. Simon, P. Sucosky and A. P. Yoganathan, Fluid mechanics of artificial heart valves, *Clin. Exp. Pharmacol. Physiol.,* 2009. 36(2): pp. 225-37.

David, T. E., C. Pollick and J. Bos, Aortic valve replacement with stentless porcine aortic bioprosthesis, *J. Thorac. Cardiovasc. Surg.,* 1990. 99(1): pp. 113-118.

David, T. E., A. Omran, G. Webb, H. Rakowski, S. Armstrong and Z. Sun, Geometric mismatch of the aortic and pulmonary roots causes aortic insufficiency after the Ross procedure, *J. Thorac. Cardiovasc. Surg.,* 1996. 112(5): pp. 1231-7.

Dellgren, G., C. M. Feindel, J. Bos, J. Ivanov and T. E. David, Aortic valve replacement with the Toronto SPV: long-term clinical and hemodynamic results, *Eur. J. Cardiothorac. Surg.,* 2002. 21: pp. 698-702.

Desai, N. D. and G. T. Christakis, Bioprosthetic Aortic Valve Replacement: Stented Pericardial and Porcine Valves, Chapter 34, Cardiac Surgery in the Adult, 2008. pp. 857-894. [Viewed from: http://cardiacsurgery.ctsnetbooks.org/cgi/content/full/3/2008/857?ck=nck] on 22/06/11.

Emery, R. W., R. W. Anderson, W. G. Lindsay C. R. Jorgensen, Y. Wang and D. M. Nicoloff, Clinical and hemodynamic results with the St. Jude medical aortic valve prosthesis, *Surg. Forum,* 1979. 30: pp. 235-238.

Emery, R. W., A. M. Emery, A. Knutsen and G. V. Raikar, Aortic Valve Replacement with a Mechanical Cardiac Valve Prosthesis, Chapter 33, Cardiac Surgery in the Adult, 2008. pp. 841-856. [Viewed from: http://cardiacsurgery.ctsnetbooks.org/cgi/content/full /3/2008/841] on 22/06/11.

Ellis, J. T., T. M. Wick and A. P. Yoganathan, Prosthesis-induced hemolysis: mechanisms and quantification of shear stress, *J. Heart. Valve Dis.,* 1998. 7(4): pp. 376-386.

Ezekowitz, M. D., Anticoagulation management of valve replacement patients, *J. Heart Valve Dis.,* 2002. 11,Suppl 1, pp. S56-60.

Gott, V. L., D. E. Alejo and D. E. Cameron, Mechanical heart valves: 50 years of evolution, *Ann.Thorac. Surg.,* 2003. 76(6): pp. S2230-2239.

Grados, A. G., M. A. D. Aguilar, C. V. Trujillo, L. E. N. O-Pinzon, J. M. Mendez, B. F. A. Rivera, E. R. Perez and H. A. Rosales, Tricuspid valve replacement in antiphospholipid syndrome patient: Anticoagulation management. Case report, *Appl. Cardiopulm. Pathophysiol.,* 2007. 11(4): pp. 64-65.

Gudbjartsson, T., S. Aranki and L. H. Cohn, Mechanical/Bioprosthetic Mitral Valve Replacement, Chapter 38, *Cardiac Surgery in the Adult,* 2003. pp. 951-986. [Viewed from:
http://cardiacsurgery.ctsnetbooks.org/cgi/content/full/2/2003/951]
on 22/06/11.

Gudbjartsson, T., T. Absi, and S. Aranki, Mitral Valve Replacement, Chapter 43, *Cardiac Surgery in the Adult,* L.H. Cohn, Editor 2008, New York: McGraw-Hill. p. 1031-1068. [Viewed from: http://cardiacsurgery.ctsnetbooks.org/cgi/content/full/3/2008/1031

Gelsomino, S., Morocutti, G., Frassani, R., Da Col, P., Carella, R., and Livi, U., Usefulness of the Cryolife O'Brien stentless suprannular aortic valve to prevent prosthesis-patient mismatch in the small aortic root. *Journal of the American College of Cardiology,* 2002. 39(11): p. 1845-1851.

Gonzalez B., H. Benitez, K. Rufino, M. Fernandez and W. Echevarria, Biomechanics of mechanical heart valve, Applications of Engineering Mechanics in Medicine, GED at University of Puerto Rico, Mayaguez, 2003.

Hanssen, O. B., P. Gjertsson, E. Houltz, B. Wranne, P. Ask, D. Lyod and K. Caidahl, Net Pressure Gradients in Aortic Prosthetic Valves can be Estimated by Doppler, *J. Am. Soc. Echocardiogra.,* 2003. 16(8): pp. 858-866.

Hammermeister, K.E, G. K Seith, W.G. Henderson,C. Oprian, T. Kim and S. Rahimtoola, A comparison of outcomes in men 11 years after heart-valve replacement with a mechanical valve or bioprosthesis, *N. Engl. J. Med.,* 1993. 328(18): pp. 1289-96.

Khan, S. S., A. Trento, M. D. Robertis, R. M. Kass, M. Sandhu, L. S. C. Czer, C. Blanche, S. Raissi, G. P. Fontana, W. Cheng, A. Chaux and J. M. Matloff, Twenty-year comparison of tissue and mechanical valve replacement, *J. Thorac. Cardiovasc. Surg.,* 2001. 122: pp. 257-269.

Kortke, H. and R. Korfer, International normalized ratio self-management after mechanical heart valve replacement: is an early start advantageous?, *Ann. Thorac. Surg.,* 2001. 72(1): pp. 44-48.

Knott, E., H. Reul, M. Knoch, U. Steinseifer and G. Rau, In vitro comparison of aortic heart valve prosthesis, *J. Thorac. Cardiovasc. Surg.,* 1988. 96(6): pp. 952-961.

Mendelson, K. and F. J. Schoen, Heart valve Tissue Engineering: Concepts, Approaches, Progress, and Chanllenges, *Ann. Biomed. Eng.,* 2006. 34(12): pp. 1799-1819.

Michell, R., R. Jonas and F. Schoen, Pathalogy of explanted cyropreserved allograft heart valves: comparison with aortic valves from orthotopic heart transplants, *J. Thorac. Cardiovasc. Surg.,* 1998.115:pp. 118-27.

Mohri, H., R. J. Nelson, M. P. Sands and K. A. Merendino, Homotransplantation of the aortic valve, *Surg. Forum,* 1966. 17: pp. 173-174.

Mohty, D., J. F. Malouf, S. E. Girard, H. V. Schaff, D. E. Grill, M. E. E. Sarano and F. A Miller Jr, Impact of prosthesis-patient mismatch on long-term survival in patients with small St Jude medical mechanical prostheses in the aortic position, *Circulation,* 2006. 113(3): pp. 420-426.

Mol, A., A. I. P. M. Smits, C. V. C. Bouten and F. P. T. Baaijjens, Tissue engineering of heart valves: advances and current challenges, *Expert Rev. Med. Devices,* 2009. 6(3): pp. 259-275.

Morsi, Y., Kogure, M. & Umezu, M. 1999, "Relative blood damage index of the jellyfish valve and the Bjork-Shiley tilting-disk valve," *Journal of Artificial Organs,* vol. 2, no. 2, pp. 163-169.

Murray, G., Aortic Valve Transplants, *Angiology,* 1960. 11(2): pp. 99-102.

Murray, G., Homologous Aortic-Valve-Segment Transplants as Surgical Treatment for Aortic and Mitral Insufficiency, *Angiology,* 1956. 7(5): pp. 466-471.

Nelson, R. J., H. Mohri, L. C. Winterscheid, D. H. Dillard and K. A. Merendino, Early clinical experience with homotransplantation of the aortic valve, *Circulation,* 1967. 35(4 Suppl).

O'Brien, M. F., E. G. Stafford, M. A. Gardner, P. G. Pohlner and D. C. McGiffin, A comparison of aortic valve replacement with viable cryopreserved and fresh allograft valves, with a note on chromosomal studies, *J. Thorac. Cardiovasc. Surg.,* 1987. 94(6): pp. 812-23.

Paneth, M. and M. F. O'Brien, Transplantation of human homograft aortic valve, *Thorax,* 1966. 21: pp. 115-117.

Pick, A., Mechanical Heart Valve Replacements—Prosthetic Heart Valve Review, The Patient's Guide to Heart Valve Surgery, 2010. [Viewed from: http://www.heart-valve-surgery.com/mechanical-prosthetic-heart-valve.php] on 22/06/11.

Rajani, R., D. Mukherjee and J. B. Chambers, Doppler echocardiography in normally functioning replacement aortic valves: A review of 129 studies, *J. Heart Valve Dis.,* 2007. 16(5): pp. 519-535.

Ross, D. N., Homograft Replacement of the Aortic Valve, *The Lancet,* 1962. 280(7254): p. 487.

Sacks, M. S., F. J. Schoen and J. E. Mayer, Bioengineering challenges for heart valve tissue engineering, *Annu. Rev. Biomed. Eng.,* 2009. 11: pp. 289-313.

Sands, M. P., R. J. Nelson, H. Mohri and K. A. Merendino, The procurement and preparation of aortic valve homografts, *Surgery,* 1967. 62(5): pp. 839-42.

Sallam, A. M. and N. H. Hwang, Human red blood cell hemolysis in a turbulent shear flow: contribution of Reynolds sheer stresses, *Biorheology,* 1984. 21(6): pp. 783-797.

Schoen, F., J. Titus, and G. Lawrie, Bioengineering aspects of heart valve replacement. *Annals of Biomedical Engineering*, 1982. 10(3): p. 97-128.

Schoen. F. J., R. L. Levy and H. R. Piehler, Pathological considerations in replacement cardiac valves, *J. Cardiovasc.Pathol.*, 1992. 1: pp. 29-52.

Schoen FJ, Levy RJ. Founder's Award, 25th Annual Meeting of the Society for Biomaterials, perspectives. Providence, RI, April 28-May 2, 1999. Tissue heart valves: current challenges and future research perspectives *J Biomed Mater Res*. 1999 Dec 15;47(4):439-65.

Shiono, M., Y. Sezai, A. Sezai, M. Hata, M Iida and N. Negishi, Long-Term Results of the Cloth-Covered Starr-Edwards Ball Valve, *Ann. Thorac. Surg.*, 2005. 80(1): pp. 204-209.

Shumacker, H. B. Jr., The evolution of cardiac surgery, 1992. [Viewed from: http://books.google.com.au/books on 22/06/11.

Sodian, R., J. S. Sperling, D. P. Martin, A. Egozy, U. Stock, J. E. Mayer Jr. and J. P. Vacanti, Fabrication of a tri-leaflet heart valve scaffold from a polyhydroxyalkanoate biopolyester for use in tissue engineering, *Tissue Eng.*, 2000. 6(2): pp. 183-188.

Suezawa, T., T. Morimoto, T. Jinno and M. Tago, Forty-Year Survival With Smeloff-Cutter and Starr-Edwards Prostheses, *Ann. Thorac. Surg.*, 2008. 85(3): pp. e14-16.

Tarzia, V., T. Bottio, L. Testolin and G. Gerosa, Extended (31 years) durability of a Starr-Edwards prosthesis in mitral position, *Interact CardioVasc. Thorac. Surg.*, 2007. 6: pp. 570-571.

Wang, C. H., T. M. Lee and C. H. Tsai, Hemodynamic Evaluation of St. Jude Medical Aortic Valve Prostheses Using Dobutamine Stress Echocardiography, *Acta Cardiol. Sin.*, 2003. 19(4): pp. 229-236.

Yacoub, M. H. and J. J. Takkenberg, Will Heart Valve Tissue Engineering Change the World? Nat. *Clin. Pract. Cardiovasc. Med.*, 2005. 2(2): pp. 60-61.

Chapter IV

Biomaterials Characterisation

4.1. Introduction

As discussed in Chapter 3, artificial heart valves suffer from serious drawbacks, as the recipients have to take anti-coagulant medicines throughout their lives to avoid calcification and thromboembolism, and they are not feasible for children in that they have to undergo multiple surgeries because of continuous growth and development. Moreover, artificial heart valves cannot synchronize with dynamic changes of the heart, and self-healing or -repairing is not feasible as these valves are not self-regenerative. Tissue-engineered heart valves, on the other hand, offer an alternative option free from all the defects like thrombogenesis and infection, and they offer long-term viability. In addition, tissue-engineered concept offers remodelling and self-growth, repair in infants and children, to eliminate the need of repeated surgical procedures (Mol et al., 2009; Knight et al., 2008; Yacoub and Takkenberg, 2005).

An Extra Cellular Matrix (ECM) is a major component of heart valves, and they respond effectively to change according to the hemodynamic forces imposed by the heart. To create a tissue-engineered heart valve, the main characteristics of ECM must be attained to ensure optimum functionality and to obtain the correct characteristics for lifetime working of the valve (Mol et al., 2009; Meldenson and Schoen, 2006; Knight et al., 2008).

In tissue-engineering concept, the materials selected should be innovative and should mimic the natural material of heart valve (Schoen, 2011). Scaffold is a three-dimensional structure that accommodates cells in their matrices and helps the growth of new cells in three dimensions to create the valve. The three-dimensional matrix is the key for the tissue-engineering development, and it should provide the environment for the cell to breed and reinstate the tissue and perform its ecological function. Therefore, material selection and optimization for the creation of the 3D scaffold is one of the most important research domains in tissue engineering (Schoen, 2011; Meldenson and Schoen, 2006). Today, biomedical materials are being used for tissue-engineering applications of cardiovascular system repair, bone fracture repair, burned skin replacement, cartilage regeneration, etc. For tissue engineering of a heart valve, various types of materials from natural origin or synthetic ones are being explored as scaffolds. Natural polymer materials are the ones produced as an end result of naturally occurring sources. These materials may be produced by chain modification of the naturally produced ingredients, whereas synthetic polymeric materials are manmade polymers.

The types of the material that are to be discussed here cover the following:

- Biological or Natural materials including:
  - Collagen
  - Gelatin
  - Alginate
  - Hyaluronic Acid
  - Chitin and Chitosan
- Synthetic materials including:
  - Polyesters
  - Polyglycolic Acid(PGA)Polylactide (PLA): and their co-polymers
  - Poly(caprolactone)
  - Polypropylene fumarates
  - Polyanhydrides
  - Polyurethanes
  - Polyetherester amide
  - Poly(ortho esters)
  - Pseudo polyamino acid
  - Polyalkyl cyanoacrylates
  - Polyphosphazenes
  - Polyphosphoester
  - Hydrogel materials

Generally, there are various advantages and disadvantages offered by the different types of materials used in the construction of the scaffold. Natural material scaffolds provide important biological information, and they have the cell recognition signals that can prevent an immune response. However, they suffer from drawbacks such as difficulty in penetrating into the interior when being developed from the exterior part, and they may produce alteration in the physical properties when decellularised along with immunologic reaction and calcification. However, biodegradable synthetic polymers are known to possess a number of improvements above natural ones. With synthetic polymers one can easily and readily alter and change mechanical properties and the rate of degradation kinetics appropriate to various engineering applications. In addition, synthetic polymers offer the advantage of being able to be fabricated into a variety of shapes with any required morphological features and chemical functionality that would favour tissue in-growth (Gunatillake and Adhikari, 2003). Moreover, scaffolds made up from synthetic materials provide advantages such as ease of reproducibility and ability to control the structure properties such as porosity (i.e., being able to absorb fluids) and degradation rate. However, they suffer from the drawback of lack of control over cell and tissue adhesion, which may cause adverse reaction in the body (Willerth and Sakiyama-Elbert, 2008; Mol et al., 2009).

Generally, however, it should be noted that there is a lack of appropriate research to define which material to be used for the type of tissue being developed. Researchers have been following a trial-and-error approach (Boccaccio et al., 2011) to determine the best suitable material or materials for the scaffold manufacturing. The researchers proposed the use of various combinations of organic and inorganic materials that best suit the application being considered. However, when it comes to manufacturing a scaffold for tissue-engineering applications, it is usually made up from a combination of organic materials of both natural and synthetic polymers. Inorganic materials are usually used to manufacture scaffolds for the development of hard tissues, such as tissue engineering of the bone, and hence they are not used for development of the soft tissues.

Nevertheless, the recent advances in the field of tissue engineering have explored new materials that are more suitable for various applications. However, the materials of natural origin derived from xenogenic or homogenic remain the same, where the main sources of materials are still cow, pig or from the donor's body. Moreover, it should be pointed out that the physical and chemical properties of tissue-engineering scaffolds can directly influence the tissue formation, and this is another factor that must be considered in the selection of materials for tissue-engineering applications. In this chapter, the two types of biological and biodegradable polymer materials are briefly presented and discussed.

4.2. Biological Materials

Biodegradable biological materials are gleaned from animal or plants, and they include collagen, gelatin, glycosaminoglycans (GAG)s, chitosan, chitin and alginate; they are widely used in various tissue-engineering applications alone and with other materials. In this section, the principle characteristics are summarized below.

4.2.1. Collagen

Collagen is an extremely dedicated protein that originates from mammals' cells and constitutes up to 35% of the body protein and presents the main element of fascia, skin, cartilage, ligaments, tendons and, bone. It has a unique structure consisting of long fibrous, which are main components of ECM with dissimilar roles to globular proteins including enzymes. Collagen holds significant tensile strength and in muscle tissue represents about 6% of the mass of the tendinous muscles; it operates as a foremost component of endomysium (Willerth and Sakiyama-Elbert, 2008; Nguyen and Lee, 2010).

Together with soft keratin, collagen is accountable for the elasticity and strength of the skin, and its deprivation results in wrinkles in older people. Moreover, it is normally found in joint tissues and is widely used in cosmetic and burn surgeries and tissue development, and it has been used to enhance cell adhesion in tissue engineering in various applications, including blood vessels and heart valves. On the negative side, collagen could transmit pathogens and induce immune reaction, and it needs to be completely purified before seeded to make it less antigenic. Moreover, it has no mechanical strength, it is difficult to control its biodegradability and it is not easy to handle and fabricate. The various types of collagens are assembled by about 20 genes and encompassed of three polypeptide chains, which contain 1,000 amino acids with the existence of glycine, Proline, and Hydroxyproline as major residues (Gelse et al., 2003). Figure 4.1 shows an illustration of the collagen as described in

Science Daily, 2006. There are up to ten kinds of diverse structures of collagen, and Table 4.1 lists only five strains (Lodish et al., 2000).

Table 4.1. Five primary kinds of collagen: Characteristics and distribution

Types	Existence	Collagen Fibrils	Chemical Disposition	Illustrative Tissues	Sources
I	300nm Three piece of helix fibril (α_1:2, α_2: 1)	67nm cross fibre	Low hydroxylysine, low carbohydrates	Derma, tendon, bone, ligament,	Fibroblast
II	300nm Three piece of helix fibril (α_1: 3)	67nm cross fibre	High hydroxylysine, High carbohydrates	Cartilage, intervertebral disk, notochord	Fibroblast, Chondroblast
III	300nm Three piece of helix fibril (α_1: 3)	67nm cross fibre	High hydroxyproline, low hydroxylysine, low carbohydrates	Derma, blood vessel	Reticular cell
V	390nm C-Globosity bottom (α_1: 3)	Reticular. Not forming fibre bundle	Higher hydroxyproline, high carbohydrates	Basement membrane	Epithelial cell, Endothelial cell
VI	390nm N-Globosity bottom	Fine fibre		Located at tissue clearance and distributed with collagen	Smooth muscle cell, Myoblast

Moreover, collagen is a kind of stringy albuminoid that cannot be dissolved in water, diluted acid and alkali. Additionally, its hydrolysis produces gelatin (Imeson, 2006). It is distributed in all parts of the body and mostly derived from fibroblast, chondroblast, osteoblast, and Schwann cells from nerve tissues. The basic component of collagen molecular is tropo-collagen, whose length is 280nm with a diameter of 15nm, and the molecular weight reaching 300,000 with various percentages of amino acid. Still, tropo-collagen is formed by three polypeptide chains and has the helix structure, and those chains are connected with each other by electrovalent and hydrogen bonds. The general practice in stabilizing the collagen fibre in tissue is to apply some exterior mechanical or chemical cross-linking methods. These methods are used to modify the physicochemical properties of the collagen, for example, reducing the solubility, water absorption, degradability and antigenicity, and to improve the mechanical property (Chevallay and Herbage, 2000).

The physical and chemical cross-linking methods that are widely used include ultraviolet irradiation, gravity dehydration and glutaraldehyde, as well as genipin. However, glutaraldehyde (GTA) still widely remains cytotoxic. Hexamethylene Diisocyanate (HMDIC), on the other hand, has double function, and its isocyanic acid group sometime could react with amino or hydroxy on collagen molecular, while 1-enthyl-3 (3-dimethylaminopropyol) – carbodiimide (EDC) is one kind of compound that has good chemical properties, has no cytotoxicity and is used to improve the biocompatibility of collagen.

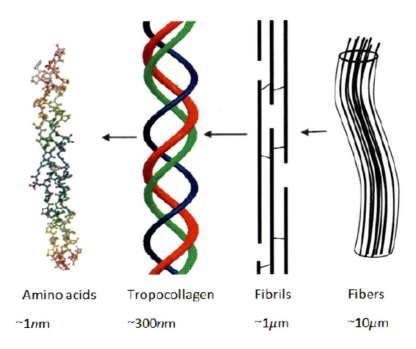

Figure 4.1. Artistic illustration of collagen Structure (Science Daily, 2006).

4.2.2. Gelatin

Gelatin is a natural biopolymer derived from partial hydrolysis of collagen and has the most identical compositions and biological properties as those of collagen; it has been used in wound dressing, drug delivery system, etc. (Imeson, 2006; Nguyen and Lee, 2010; McNulty, 2011). Moreover, gelatin of type A, which can be prepared from acid-treated collagens, has comparable biocompatibility property of collagen, it is inexpensive to produce and it can advance water absorption and react well with –OH groups. Additionally, gelatin is an aqueous polymer, i.e., dissolvable in water and has been applied widely in tissue engineering applications (Nguyen et al., 2010). It contains a large number of glycine (almost one in three residues, organizedevery third residue), proline and 4-hydroxyproline residues. A conventional organization is: Ala-Gly-Pro-Arg-Gly-Glu-4Hyp-Gly-Pro, and the process of gelatin fabrication is selection, cracking, degreasing, classification, pickling, neutralization, distillation, concentration, and drying (Suprakas and Mosto, 2005 ; Basavaraju et al., 2006, Shubhra et al., 2011, Vasilenko et al., 2010).

4.2.3. Alginate

Alginate is derived from algae; it is one kind of polysaccharide, and it is the compound of D-mannuronic acid and L-guluronic. Usually, there are three types: pure D-mannuronic acid, pure L-guluronic, D-mannuronic acid and L-guluronic. However, due to the structure of the chain of the alginate, the properties of different fibres are dissimilar. Sylvine and sodium salt of alginate can be dissolved in water, where other salt cannot be dissolved (Qin, 2003; Kakita and Kamishima, 2009; Jejurikar et al., 2010). The chemical structure of alginate is illustrated in Figure 4.2.

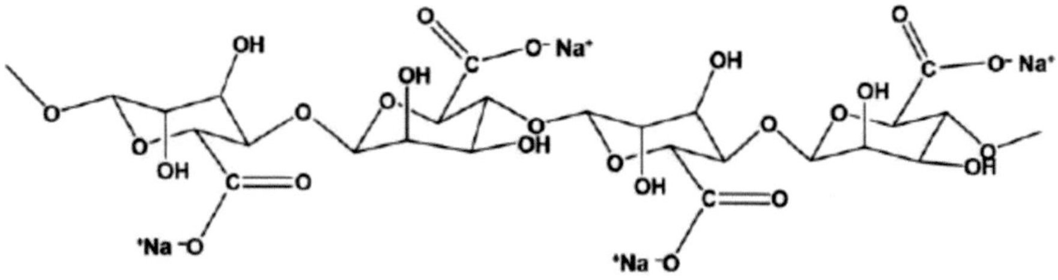

Figure 4.2. Chemical structure of alginate.

Moreover, with the existence of divalent ion, alginate can form hydrogel through ion exchange, and this kind of hydrogel has good plasticity and can be processed with different shapes. Still, alginate can degrade by enzymolysis *in vivo* with no cytotoxicity, and similar to gelatine, alginate has been used widely for wound dressing, drug delivery, cell culture media and neural tissue-engineering applications (Willerth and Sakiyama-Elbert, 2008).

4.2.4. Hyaluronic Acid (HA)

It is stated that the designation of hyaluronic acid (HA) was created for the polysaccharide from hyalos, signifying glassy and vitreous, and uronic acid. Hyaluronic acid is an unbranched polysaccharide of replicating disaccharides comprising of D-glucuronic acid and N-acetyl-D-glucosamine, has an excellent water-retaining capacity and has visco-elastic properties. HA has been regularly used in many biomedical devices and is a vital part of the extracellular matrix, whose physiological purposes are, expressed both as the substance by itself and/or when it is being linked to various proteins. HA has a wide dissemination in almost all animal organs, different sources of HA have no species differences. The physical chemistry of HA is less proven if it is planted with other biopolymers, e.g., nucleic acids and proteins (Hargittai and Hargittai, 2008; Vázquez et al., 2010; Cushion Joints with Hyaluronic Acid, 2011).

The manufacturing process of HA is normally classified into two kinds: the extraction from animal organs and bacterial fermentation (Patil et al., 2011). The bonds, which have been established explicitly, are illustrated in Figure 4.3.

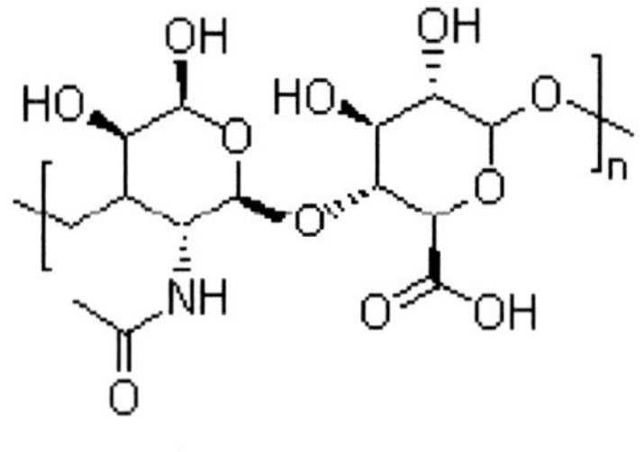

Figure 4.3. Chemical structure of Hyaluronic acid.

4.2.5. Chitin and Chitosan

Chitin, **poly (β-(1-4)-N-acetyl-D-glucosamine)**, was first identified in 1884, and is a natural polysaccharide of key value (Rinaudo, 2006; Rinaudo, 2008). This biopolymer is the most ample material after cellulose and is synthesized by an enormous number of living organisms (Jayakumar et al., 2010). Chitin can be found in nature and possesses well-organized crystalline micro-fibrils forming physical structure in the cell walls of fungi and yeast and in exoskeleton of arthropods (Rinaudo, 2006; Tamura, 2008; Soares, 2009). Moreover, Chitin is also moulded by a various other active organisms in the lower plants and animals (Kumirska et al., 2011) and has been used to enhance mechanical strength of polymer and biocompatibility. It is important for healing of wounds; it is usually derived from shell animals, and degradation product *in vivo* is oligosaccharide, therefore products are considered safe and innocuous.

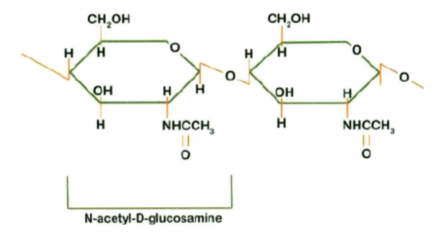

(a) (Research, viewed on 11/11/11 from
http://www.ceoe.udel.edu/horseshoecrab/research/chitin.html).

(b) (SWICOFIL, viewed on 11/11/11 from
http://www.swicofil.com/products/055chitosan.html).

Figure 4.4. Comparison of different chemical structures (a) Chitin (b) Chitosan.

Depending on the type of polymer used, once the percentage of deacetylation of chitin gets to approximately 50%, chitin transformed to chitosan, which is soluble in aqueous acidic media (Willerth and Sakiyama-Elbert, 2008; Rau, 2009; Qin, 2003; Wang et al., 2011). The process of solubilization arises as a result of the action of protonation of the –NH2 function on the C-2 position of the D-glucosamine recurrence unit, when the polysaccharide is transformed to a polyelectrolyte in acidic media It should be noted that chitosan is the only pseudo-natural cationic polymer, and, therefore, it could be applied in many situations such as flocculants for protein recovery or depollution and so on. Due to the fact that Chitosan is aqueous solutions, is fundamentally utilized as solutions, gels, or films and fibres. The principal stage in characterizing chitosan is to initially purify it: it is dissolved in excess acid and filtered on porous films made of various pore sizes up to 0.45 μm. Regulating the pH of the chemical solution to ca. 7.5mg by adding NaOH or NH_4OH triggers flocculation due to deprotonation and the insolubility of the polymer at neutral pH. The sample is subsequently cleaned with water and dried (Rinaudo, 2006; Soares, 2009).

4.2.6. Compound of Biological Materials

Biological materials all have excellent biocompatibility, but they have different mechanical properties and workability. In order to obtain high performance scaffold, researchers often combine different kinds of biological and synthetic polymers. For example chitosan, gelatin and HA could be used to fabricate bionic tissue-engineering scaffold, which would possess excellent water absorption, flexibility and biocompatibility.

4.3. Synthetic Polymeric Materials

4.3.1. Overview

The main feature of the synthetic materials is manmade polymer, which possesses a number of advantages compared to the natural ones. They are more flexible and can be constructed into various shapes and sizes (Leng and Lau, 2011; Willerth and Sakiyama-Elbert, 2008; Gunatillake and Adhikari, 2003). Moreover, with optimization of chemical composition of the macromolecule of the polymers, a specific physical and chemical, mechanical and degradation characteristics can be obtained. Still, polymers could be further self-cross-linked or cross-linked with peptides or other bioactive molecules to suit various tissue-engineering applications. Synthetic polymers are normally degraded by the simple mechanism of hydrolysis, and their rate of degradation is constant regardless of the surrender, except if there are external factors that influence the native pH deviations such as inflammations and graft degradation and so on (Leng and Lau, 2011; Cheung, 2007).

Table 4.2. Summary of the main characteristics and applications of biodegradable biological materials

Material	Authors	Main Features	Drawbacks	Possible uses and Applications
Collagen	King, 1969, Lee et al., 2001, Imeson, 2006, Willerth and Sakiyama-Elbert, 2008, Nguyen et al., 2010, Gram*, 2011	Performs as connective tissue and can strength blood vessel. Cross-linking with Carbodiimide is possible to improve biocompatibility.	Poor mechanical property. Difficult to fabricate and to control its biodegradability.	Hydrolyzed collagen can be used as a protein, Tissue engineering, Drug delivery system, Reconstructive surgery, Orthopaedics and dentistry, Ocular Surgery.
Gelatin	Basavaraju et al., 2006, Rohanizadeh et al., 2008 Nguyen et al., 2010, McNulty, 2011, Vasilenko et al., 2010, Shubhra et al., 2011	Good biocompatibility. Cost effective. Improvement of water absorption is possible Dissolvable in water.	Needs effective mechanism to control biodegradation.	Gelatin sponge used in: Wound dressing, Drug delivery system, Scaffold to support osteoblasts, Cartilage cell growth, promote bone regeneration.
Alginate	McHugh, 1987, Qin, 2003, Willerth and Sakiyama-Elbert, 2008, Kakita and Kamishima, 2009, Jejurikar et al., 2010, Bai et al., 2011,	Can be utilized to form hydrogels. Excellent plasticity. Easy to manufacture and form. Degrade by enzymolysis *in vivo* with no cytotoxicity. Large brown seaweeds are likely to be the main roots for alginate.	Less effective in cell adhesion and proliferation.	Wound dressing. Drug delivery system. Neural TE. Currently being explored in heart TE.
Hyaluronic Acid (HA)	Hargittai and Hargittai, 2008, Vázquez et al., 2010, Patil et al., 2011 *HYALURONIC ACID, 2011**	Available in high concentration in liquids in the eyes and joints. Key component of (ECM). Can be extracted from animal organs and bacterial fermentation	Less developed chemical.	Viscoelastic features are beneficial for medical applications. Lip filler and skin. Prevent the effects of aging Has a good water-maintaining ability.
Chitin & Chitosan	Qin, 2003, Moffat and Marra, 2004, Rinaudo, 2006, Rinaudo, 2008, Willerth and Sakiyama-Elbert, 2008, Tamura, 2008, Soares, 2009, Jayakumar et al., 2010, Kumirska et al., 2011, Wang et al., 2011,	Chitin is the second most ample polymer after cellulose. Can be synthesized by a number of living organisms. Chitosan is the single pseudo natural cationic polymer.	Purification is necessary for Chitosan.	Chitin can be used as substitutions for Bone, cartilage, arteries, veins, and musculo-fascial. Could reduce serum cholesterol levels. Excellent for cosmetic artefacts. Chitosan valuable for hemostatic, bacteriostatic, and spermicidal agent.

* http://EzineArticles.com/3075092.

**http://www.webmd.com/vitamins-supplements/ingredientmono-1062
   HYALURONIC%20ACID.aspx?activeIngredientId=1062&activeIngredientName=HYALURONIC%20ACID?.

Biodegradable polymers such as poly(glycolic acid), poly(lactic acid) and their copolymers, poly)p-dioxanone), and copolymers of trimethylene carbonate and glycolide have been used in various clinical such sutures, drug delivery and orthopaedic fixation implants including pins, rods and screws. However, the most extensively used synthetic polymers for tissue-engineering applications include poly (glycolic acid) (PGA), poly(lactic acid) (PLA) and their co-polymers; polycaprolactone (PCL) and polyethylene glycol (PEG) (Shalaby and Burg, 2004).

4.3.2. Major Classes of Degradable Polymers

4.3.2.1. Poly (α-esters)

While many synthetic polymers are good candidates for medical device applications, the polyesters have been particularly suitable due to their ease of degradation by hydrolysis of ester link, and the degradation products themselves are resorbable through the metallic pathways in some instances, as well as their potential to change the chemical structures to obtain different rates of degradations. Poly(α-esters) contingent on the monomeric elements could be constructed from various monomers through ring opening and condensation polymerization directions and is the initial class of biodegradable polymers and the most researched and utilized one. Moreover, Poly(α-esters)s possess significant multiplicity and synthetic versatility.

In addition, polyesters have been proposed as a candidate for the development of tissue-engineering applications, especially in the area of bone tissue engineering. Poly (α-ester)s are thermoplastic polymers with hydrolytically labile aliphatic ester linkages in their backbone (Gunatillake and Adhikari, 2003 and Domb et al., 2011). Since esterification is a chemically reversible process, all known polyesters are ideally degradable, however the degradation rate can be totally controlled over a specific time only in aliphatic polyesters, as these polymers have practically short aliphatic chains between ester bonds (Davachi et al., 2011; Bencherif et al., 2009; Nair and Laurenchin, 2007).

Biodegradable polymers belonging to polyester family are the most investigated, so far. The bulk of the published material on biodegradable polyesters belong to poly(α- hydroxyl acids) such as poly(glycolic acid) (PGA), poly(lactic acid) (PLA), and a range of their copolymers, which have an extensive history of use as synthetic biodegradable materials in a wide range of clinical applications. Some of the uses of these polymers have been put to include sutures, plates and fixtures for fracture-remedial devices as well as scaffolds for tissue culture (Khon and Langer, 1997; Gunatillake and Adhikari, 2003).

4.3.2.2. Poly (Glycolic Acid), Poly(Lactic Acid) and Their Copolymers

Poly(glycolic acid) (PGA), also called Polyglycolide, is a highly crystalline (46-50% of crystallinity), rigid thermoplastic polymer. Due to its high crystallinity, PGA is insoluble in most solvents, with the exception of highly fluorinated organic solvents like hexafluoro isopropanol, which has a glass transition temperature of 36 °C and a melting point of 225 °C. The chemical structure of PGA is given in Figure 4.5.

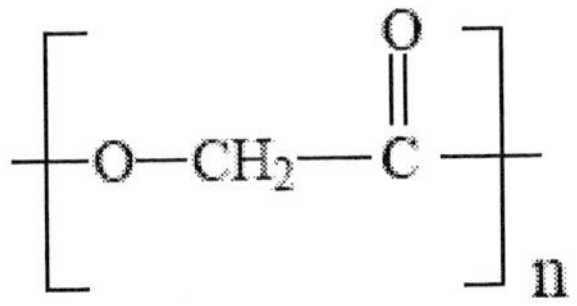

$$\left[\!\!-O\!-\!CH_2\!-\!\overset{\displaystyle\overset{O}{\|}}{C}\!-\!\right]_n$$

Figure 4.5. Chemical structure of Poly(glycolic acid) (PGA).

While normal processing techniques like extrusion, injection moulding and compression moulding may be used for the fabrication of PGA into the desired forms, care must be taken to ensure proper control of the processing conditions in view of this polymer's sensitivity to hydrolytic degradation. Although porous scaffolds and foams can also be constructed of PGA, but the type of processing technique employed can significantly affect its properties and degradation characteristics. PGA-based implants may be also fabricated by such techniques as solvent casting, particular leaching method and compression moulding (Jen et al., 1999; Mikos and Temenoff, 2000 and Gunatillake and Adhikari, 2003).

PGA degradation mechanism via homogenous erosion has been established through extensive studies (Shalaby and Burgh, 2004). It undergoes a two-stage degradation process; the first stage consists of the diffusion of water into amorphous regions and ester hydrolysis resulting in chain-scission. In the second stage, the crystalline regions that have now become predominant due to the fact that erosion of bulk of the amorphous regions are mainly involved (Gunatillake and Adhikari, 2003). Subsequently, owing to its degradation product glycolic acid being a natural metabolite, PGA finds approval as a biodegradable polymer in medical applications such as resorbable sutures (Dexon, American Cyanamide Co.).

Moreover, polyglycolide, due to its high crystallinity, possesses good mechanical properties with a young modulus of nearly 12.5 GPa, and a self-possessed polyglycolide is harder than any other known polymers; hence it has been explored as bone internal fixation devices (Features, 2011).

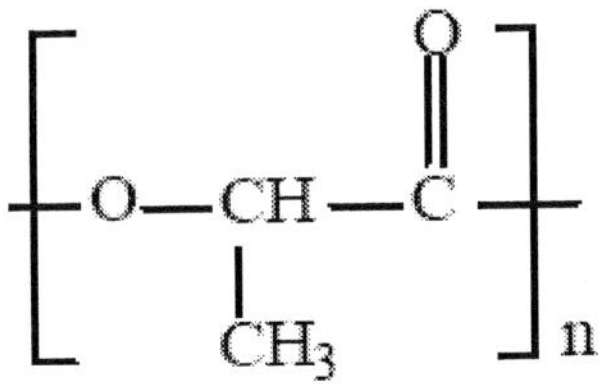

$$\left[\!\!-O\!-\!\underset{\underset{\displaystyle CH_3}{|}}{CH}\!-\!\overset{\displaystyle\overset{O}{\|}}{C}\!-\!\right]_n$$

Figure 4.6. Chemical structure of poly(lactic acid) (PLA).

Poly(lactic acid) (PLA) is extensively used alpha polyester in medical applications. Its use for implantation in human body applications has been approved by Food and Drug Administration (FDA). The chemical structure of PLA is given in Figure 4.6. (Gunatillake and Adhikari, 2003). The hydrophobic character of PLA is responsible for its slower degradation than PGA, as hydrophobicity limits water absorption of thin films and slows down the hydrolysis rate of the polymer backbone. Current research on the rate of degradation suggests that a complete degradation can occur between 12 and 24 months.

Depending on the choice of isomer used in the polymer, relatively low tensile strength and modulus of elasticity may be achieved (Cheung et al., 2007; Lau, 2007).

PLA is more hydrophobic and resilient to hydrolytic effect than PGA. Some of the biodegradable polymers approved by the Food and Drug Administration (FDA) include PLA, PGA and poly (lactic-glycolic acid) (PLGA) copolymers, illustrated in Figure 4.7 (Gunatillake and Adhikari, 2003).

$$\left[\begin{array}{c} O-CH_2-\overset{\overset{\displaystyle O}{\|}}{C}-O-CH-\overset{\overset{\displaystyle O}{\|}}{C} \\ \underset{CH_3}{|} \end{array}\right]_n$$

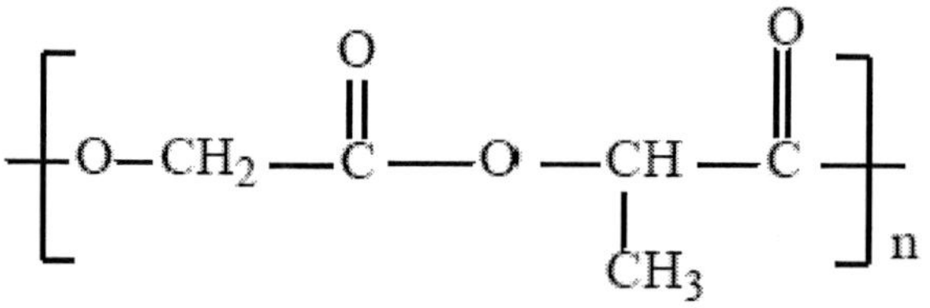

Figure 4.7. Chemical structure of Poly(glycolic-co-lactic acid).

With respect to biodegradation and biocompatibility of polylactides, it should be noted that polylactide is a biodegradable polymer. Many factors determine the actual rate of degradation. They include crystallinity, structure configuration, morphology, molecular weight, copolymer ratio, presence and amount of residual monomer, porosity, stress factors, as well as implantation site. Moreover, it has several other mechanisms through which they degrade, including thermal degradation, hydrolytic degradation and photo-degradation, which are accompanied by by-products such as lactic acid, carbon-dioxide and water. It has been an attractive commodity thermoplastic by virtue of its excellent mechanical properties, low toxicity and its ability to degrade. Moreover, various applications have been carried out to ascertain its viability, particularly in biomedical and pharmaceutical applications (Chu and Liu, 2008). Additionally, optimization of the rate of degradation of PLA, PGA and PLA/PGA copolymers could be achieved through the random hydrolysis of their ester bonds.

The first step of PLA degradation involves the formation of lactic acid, which is predominant in the human body. Still, some esterase-active enzymes are known to be capable of breaking down PGA; and glycolic acid can be excreted via urine (Vert et al., 1984; William and Mort, 1997; Gunatillake and Adhikari, 2003). The biocompatibility of PLA and PGA has been extensively investigated through both *in vitro* and *in vivo* studies. While most studies have shown they (PLA and PGA) are fairly biocompatible, some have indicated otherwise. In fact, some studies have demonstrated that porous PLA-PGA scaffolds may produce considerable systemic or local reactions or even help create unfavourable responses during the process of tissue repair. However, PLA-PGA copolymers employed in bone repair applications have been shown to be biocompatible, non-toxic and non-inflammatory. Still, the success of PLA-PGA in clinical use as sutures leads one to consider their application in fixation devices or replacement implants in musculoskeletal tissues (Nelson et al., 1977; Hollinger, 1983, Schakenraad et al., 1989, VanSliedregt et al., 1992; Verheyen et al., 1993, Gunatillake and Adhikari, 2003).

4.3.2.3. Poly(Caprolactone)

Poly(caprolactone) (PCL) is the most extensively studied polymer in the polylactone family. PCL is essentially a non-toxic polymer and is extremely tissue compatible. Its chemical structure is given in Figure 4.8. PCL has a semi-crystalline structure, with a relatively low melting temperature, in the range of 59 to 64°C, and a glass transition temperature of about –60°C. PCL has a lower degradation than PLA, is compatible with many polymers and is widely used as a base polymer in the development of drug delivery and long-term implantable devices (Gunatillake and Adhikari, 2003). The synthesis of poly (caprolactone) involves the mechanism of ring-opening polymerization, of ε-caprolactone. The reaction is initiated by low molecular weight alcohols, which helps to control the molecular weight of the final polymer, and catalysed by materials such as stannous octoate (In't Veld et al., 1997; Storey and Taylor, 1998 Gunatillake and Adhikari, 2003, Lahcini et al., 2011).

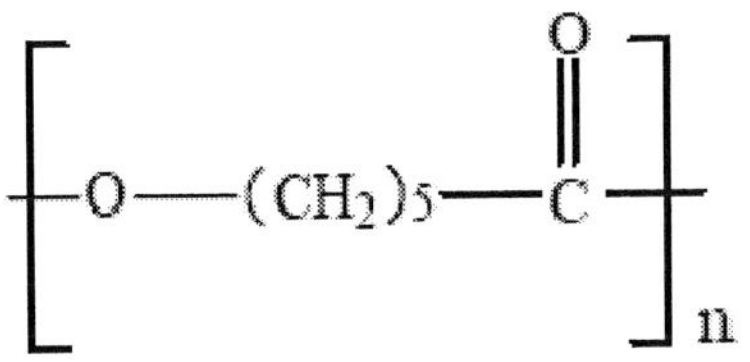

Figure 4.8. Chemical structure of PCL.

The findings from literature indicate that the degradation phase for a polylactone homopolymer is between two to three years. Still, *in vitro* trials indicated that PCL with an initial regular molecular weight of 50,000 degraded entirely after three years. Moreover copolymerisation with other lactones can greatly change the rate of degradation. For instance, a copolymer of caprolactone and valerolactone and copolymers of ε- caprolactone with dl-lactide, found to degrade more rapidly than others types of polymers (Gunatillake and Adhikari, 2003).

4.3.2.4. Poly(Propylene Fumarates)

In recent years, biodegradable-polyesters based fumaric acid became very popular among researchers in the fields. Of particular interest in this group is the copolyester poly(propylene fumarate) (PPF), which has been studied extensively. The chemical structure of PPF is given in Figure 4.9.

When this copolymer degrades, it forms fumaric acid and 1, 2-propane diol. Fumaric acid occurs in nature and is also present in the tri-carboxylic acid cycle (Krebs cycle), while 1, 2-propane diol finds use as regular diluent in drug formulations. The unsaturated sites available in the backbone of the copolymer can be exploited for further cross-linking reactions. While PPF undergoes bulk degradation, its rate of degradation is governed by the polymer structure as well as other components. The hydrolytic degradation of PPF yields fumaric acid and propylene glycol (Peter et al., 1998, Gunatillake and Adhikari, 2003).

Moreover, it was discovered in the course of *in vitro* studies that the time required to reach 20% loss in original weight ranged from 84 days for PPF/β-TCP composite to over 200 days for PPF/CaSO4 composite. The presence of β-TCP in these compositions not only facilitated an increase in mechanical strength but also provided buffering effects in controlling pH changes and keeping them to a minimum during the degradation process. While PPF does not show any long-term, detrimental, inflammatory response during subcutaneous implantation in rats, a mild inflammatory response was, however, observed initially as well as the formation of a fibrous capsule around the implant at 12 weeks (Peter et al., 1998; Gunatillake and Adhikari, 2003).

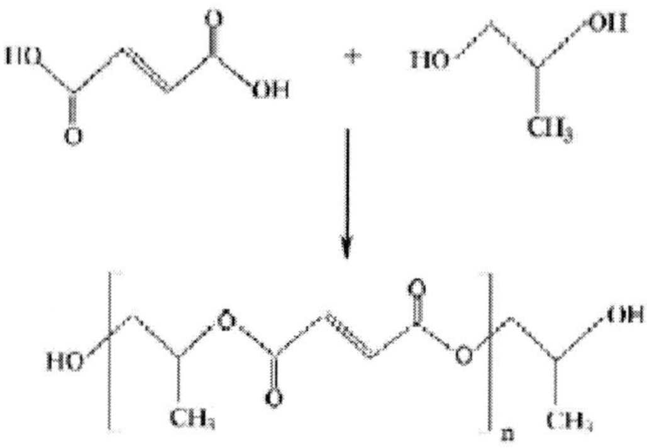

Figure 4.9. Chemical structure of Poly (propylene fumarates) (PPF).

It should be noted that PPF- based polymer networks throughout the early stages of degradation characteristically display an increase in their mechanical properties. Though due to the conflicting processes of hydrolytic degradation and thermal induced cross-linking, PPF-based networks show biphasic degradation performance (Timmer et al., 2003(A); Timmer et al., 2003(B)). It is a vital biodegradable and cross-linkable polymer used for bone tissue engineering (Wang et al., 2006).

4.3.2.5. Polyanhydrides

Polyanhydrides have sustained biocompatibility, and their degradation mechanism is based on surface erosion. Owing to their hydrolytic instability and surface eroding nature, these polymers are widely proposed as a good candidate for controlled drug delivery (Timmer et al., 2003). The dehydration of the corresponding diacid or a mixture of diacids produces Polyanhydrides, by melt polycondensation process. The monomeric dicarboxylic acids are converted to the mixed anhydride of acetic acid by refluxing with excess acetic anhydride. For the synthesis of high molecular weight polymers, a process of melt condensation of pre-polymer in vacuum under nitrogen sweep is normally applied. The chemical structure of these polymers is shown in Figure 4.10 (Gunatillake and Adhikari, 2003, Nair and Laurencin, 2007; Sabir et al., 2009).

Moreover, the literature suggests that to synthesize polyanhydrides, one can either apply melt condensation of diacid esters, ROP (Ring Open Polymerization) of anhydrides, interfacial condensation, and dehydrochlorination of diacids and diacid chlorides or can obtain synthesis by the utilization of the response of diacyl chlorides with blending substances for instance phosgene or diphosgene (Domb and Nudelman,1995; Kumar et al., 2002; Nair and Laurecin, 2006; Pielichowska and Blazewicz, 2010). Polyanhydrides, owing to their exceedingly sensitive aliphatic anhydride bonds, are considered to be one of most hydrolytically labile type of polymers. In general, different kinds of catalyst structures for polymerization can be identified, which allowed the production of high molecular weight polyanhydrides.

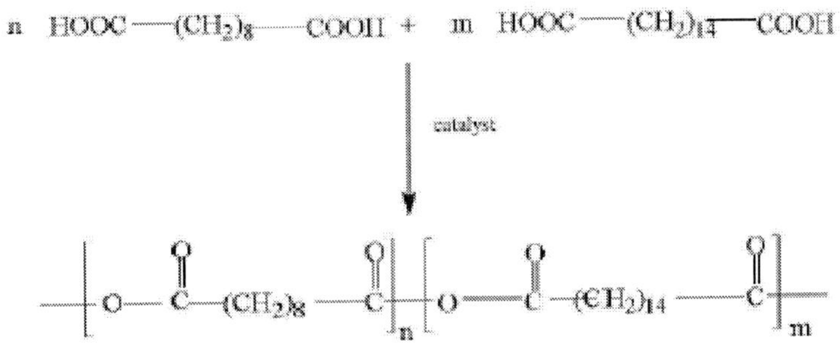

$$n \ \ HOOC-(CH_2)_8-COOH + \ m \ \ HOOC-(CH_2)_{14}-COOH$$

catalyst

Figure 4.10. Chemical structure of Polyanhydrides.

Both homo- and co-polyanhydrides display various characteristics and have been constructed by the melt condensation technique. However, it should be noted that due to fast rate of degradation of aliphatic homo-polyanhydrides, including poly(sebacic anhydride) (PSA), their applications are constrained. (Ulery et al., 2011). With the aim of developing polymers types with good degree of controllable rate of degradation rate and construction, many co-polymers of sebacic anhydride and hydrophobic aromatic co-monomers have been explored (Nair and Laurencin, 2007; Vasheghani-Farahani and Khorram, 2002; Gopferich and Tessmar, 2002). Moreover, Petersen et al. (2011) have introduced a new class of biodegradable, amphiphilic polyanhydrides containing suitable protein and cell material. These materials offer challenges for the design of functional tissue having a proper cellular integration to eliminate bacterial infections and subtle inflammatory response (Petersen et al., 2011).

Owing to polyanhydrides poor mechanical performance, they are not preferably used for load-bearing applications, such as orthopaedic implants, but these type of polymers are widely proposed as ideal candidates for drug delivery applications (Nair and Laurencin, 2007; Cui et al., 2004; Zhang et al., 2002). This motivated the researchers to develop a new class of polymers known as poly(anhydride-co-imides), which possess imide segment in the polymer backbone imparting unusual strength still with surface eroding properties. The osteocompatibility of these polymers were examined using a rat tibial model, and the results showed that the defects without any modification restored in 12 days but the defects that were treated with poly(anhydride-co-imides) generated endosteal bone growth after three days of the treatment. Moreover, after 30 days the bridges of cortical bone around the implanted matrices were observed, indicating the good osteocompatibility of the matrices (Nair and Laurencin, 2007).

With cross-linked polyanhydrides, there are a number of methods used for cross-linked; for example, in the study of Ibim et al. 1998 and Muggli et al. (1999), robust surface-eroding polymers were constructed from methacrylated anhydride monomers of sebacic acid (MSA) and 1, 6-bis(carboxyphenoxy) hexane (MCPH). Subsequently, these multifunctional monomers were photopolymerized using ultraviolet light to fabricate approvingly cross-linked polyanhydride structure (Muggli et al., 1999; Ibim et al., 1998). Another approach was explored with the aim of increasing the strength of polyanhydrides, by introducing acrylic functional groups in the monomeric units to construct injectable photo-cross-linkable polyanhydrides. It has been proposed that injectable anhydrides could be used for stuffing brokenly shaped bone or for soft tissue treatments that necessitate substance of a liquid or putty-like make up, that could be mold into a preferred geometry (Moffat and Marra, 2004; Muggli et al., 1999).

4.3.2.6. Polyurethane

Polyurethane polymers possess excellent biocompatibility and mechanical properties and represent the type of polymers that are considered to be the core class of synthetic elastomers. They have been developed for various medical devices, such as cardiac pacemakers and vascular grafts and even heart valves (Baroli, 2006; Gunatillake and Adhikari, 2003). Polyurethanes are generally formulated by the poly-condensation reaction of diisocyanates with alcohols and amines (Dey et al., 2008) and due to the symmetrical structure of the diisocyanate, the polyurethane is semicrystalline and possesses good mechanical properties (Wu et al., 2001). Moreover, due to their favorable biological and mechanical characteristics, attempts were made to develop biodegradable polyurethanes. However, due to the toxicity of common diisocyanates such as 4,4'-methylenediphenyl diisocyanate (MDI) and toluene diisocyanate (TDI), other aliphatic diisocyanates have been proposed for the construction of biodegradable polyurethanes (Gunatillake and Adhikari, 2003; Nair and Laurencin, 2007)).

Still, it should be noted that Lysine diisocyanate (LDI), and 1,4-diisocyanatobutane (BDI) are a few degradable poly(ester urethanes) that were formulated by reacting LDI with polyester diols or triols based on D,L-lactide, caprolactone and other co-polymers. An important factor is that functioning moieties such as ascorbic acid and glucose can be amalgamated into these peptide-based polymer configurations, making them very attractive in encouraging cell attachment, differentiation and proliferation with no harmful effects (Nair and Laurencin, 2007).

Moreover, it is well recognised that the use of injectable biodegradable polymers is very attractive as they can eliminate various limitations associated with the existing surgical practices and constructed implants. To date, various biodegradable injectable hydrogels constructs have been proposed; however, due to poor mechanical properties and controlled degradability, only a limited numbers of studies are related to the formulation of injectable materials appropriated for orthopaedic applications (Nair and Laurencin, 2007). This encouraged researchers to explore the use of based polyurethane polymer as an injectable biodegradable material. Subsequently, the group at CSIRO Melbourne Australia led by DR Gunatillake constructed a distinctive injectable, two-component LDI-based polyurethane structure that heals *in situ* for orthopaedic applications. It was argued that such polymer can be dispensed arthroscopically in liquid form and polymerizes at functional temperature *in situ* to supply the required bonding robustness and mechanical integrity better than the extensively utilized bone cements. Later, it was commercialised under the name of PolyNova® (Gunatillake and Adhikari, 2003). Moreover, it has been demonstrated that such polyurethane polymers appear to encourage cell adhesion and proliferation (Vlad et al., 2010; Nair and Laurencin, 2007). However, it has been argued that the use of polyurethane-based hydrogels for biomedical applications could be attractive as they are biocompatible and possess robust hydrogen bonds that could provide superior mechanical properties. In addition, they can be used for targeted drug delivery and tissue-engineering applications (Patel and Mequanint, 2009; Bonzani et al., 2007).

4.3.2.7. Poly (Ether Ester Amide)

These co-polymers possess excellent thermal and mechanical properties, and they are mainly fabricated by polycondensation of PEG diester-diamide, which acts as a hydrophobic system. Diester-diamide can be used as a hydrophobic mass for generating adjustable physical cross-links that embrace hard-wearing mechanical properties in the swollen condition. In this circumstance, PEG is the "soft" portion and diester-diamide performs as the "hard" portion (Martina and Hutmacher, 2007). Still, it is worth noting that the rate of degradation of poly (ether ester amide) is initiated as a result of the hydrolytic cleavage of the ester or ether bonds without effecting the integrity of the amide segments (Patel and Mequanint, 2008; Nair and Laurencin, 2007). Various efforts were also carried out to control and to accelerate the rate of degradation of poly (ether ester amide) by, for example, mixing amino acid elements in the polymer backbone. Still, for the applications of site definite delivery of small hydrophobic drugs and peptides a development is still underway since 2007, to advance a poly (ester amide) blend (CAMEO®) for these purposes (Nair and Laurencin, 2007).

4.3.2.8. Poly (Ortho Esters)

It should be noted that with poly (ortho esters), by utilizing diols with different degrees of chain configurations, one can control degradation rate and glass transition temperatures and pH sensitivity. ALZA Corporation (Alzamer®) aimed at addressing the shortcomings of bulk eroding biodegradable polymers and developed polies (ortho esters) as hydrophobic, hydrolytically susceptible backbones, surface eroding polymer for drug delivery. Still, the pH sensitivity of the poly (ortho esters) makes them a suitable candidate for many drug delivery applications; in addition, the rate of drug release is principally controlled by the hydrolysis via the use of acidic or elementary excipients (Leadley et al., 1998). Currently, there are four different classes of poly (ortho esters) that have been developed (Ulery et al., 2011; Nair and Laurencin, 2007).

Figure 4.11 illustrates the structures of different types of poly(ortho esters), and of these, only POE IV has demonstrated enough capabilities for possible commercialization. Moreover, it should be noted that this type of polymer is stable at room temperature when stored under the correct (anhydrous) surroundings. In addition, the erosion progression is restricted principally to the surface layers, and, more importantly, the polymer has a good synthesis versatility to create polymers of the required mechanical and thermal properties with the preferred erosion rates (Heller et al., 2002; Nair and Laurencin, 2007; Paesen et al., 2007).

4.3.2.9. Pseudo Poly (Amino Acid)

Synthetic biodegradable polymers, using natural metabolites as monomers, have been proposed as an effective class of biomaterials. However, though poly (amino acid)s are ubiquitous and naturally-occurring biodegradable polymers; they suffer from immunogenicity and poor mechanical performances and as such limit their medical applications (Nair and Laurencin, 2007; Heller et al., 2002).

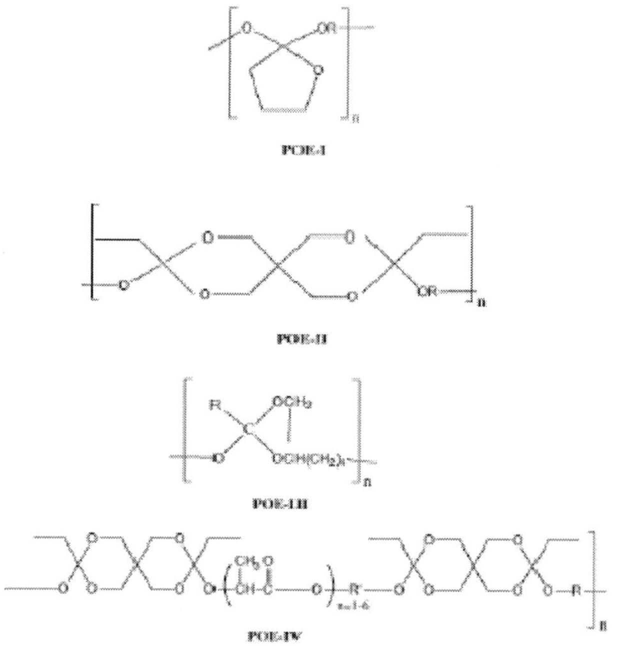

Figure 4.11. Structures of different poly (ortho ester)s.

The biodegradation of such polymers into the corresponding naturally metabolizabled monomers and their derivatives renders the polymers biocompatible. Amino acid "monomers" seem a logical choice for the development of such biomaterials. Despite their biocompatibility, use of poly (amino acids) is limited by practical difficulties like insolubility in common organic solvents, unpredictable water intake and swelling behaviour, etc., which have been traced back to the highly crystalline structure and hydrogen bonding induced by the sequence of amide (peptide) bonds in the polymer backbone. Henceforth, introduction of non-amide bonds alternating with the amide (peptide) link in the poly (amino acid) backbone is being investigated as one of the ways to elude such properties. The resulting polymer would be called a "pseudo"poly(amino acid) and the non-peptide link is expected to impart properties that are potentially favourable for various biomaterial applications.

However, note that the tyrosine-derived poly(amino acids) are considered to be one of the highly searched type of polymer (Nair and Laurencin, 2007). Starting from L-tyrosine and its deaminated analogue, 3-(4-para-hydroxy)-phenyl propionic acid, a diphenolic structure containing an amide linkage was synthesized following standard procedures of peptide synthesis. Moreover, due to aromatic backbone, these polymers demonstrate good mechanical properties and therefore could function as a biodegradable polymer for various biomedical applications.

4.3.2.10. Poly (Alkyl Cyanoacrylates)

Poly (alkyl cyanoacrylates)s (PACA), first developed 25 years ago, have good *in vivo* degradation potential and good acceptance by living tissues (Vauthier et al., 2007). This type of polymers have, to date, been investigated as excellent synthetic surgical glue, skin adhesive and an embolic material. Moreover, recently, poly (alkyl cyanoacrylate)s were considered to be one of the first biodegradable polymers used for developing nanoparticles for drug delivery applications (Nair and Laurencin, 2007; Paesen et al., 2007; Vauthier et al., 2007).

Figure 4.12 illustrates the chemical structure of PACA.

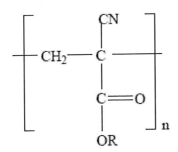

Figure 4.12. Chemical structure of poly (alkyl cyano acrylate).

The rate of degradation of poly (cyano acrylates) primarily functions on the length of the alkyl side groups. They possess quickest degradation rate among other polymers. Although the lower alkyl derivatives, such as poly (methyl cyano acrylate), degrade in an aqueous environment within few hours, its degradation can lead to toxic by-products such as cyanoacetic acid and formaldehyde (Gupta and Lopina, 2002). However, owing to the quick rate of polymerization, these monomers have been developed as tissue-bonding agents for skin applications and are used under the commercial name of Dermabonds® (2-octyl cyanoacrylate. Moreover, the special feature of hydrophobic interactions with oligodeoxynucleotides (ODN) qualify, poly(alkyl cyanoacrylates) as good candidate for gene delivery vehicles (Nair and Laurencin, 2007; Vauthier et al., 2007).

4.3.2.11. Polyphosphazenes

Different organic, inorganic and or inorganic-organic hybrid polymers have been explored as possible biodegradable biomaterials for various tissue engineering applications. Polyphosphazenes types of polymers are considered as hybrid ones with a backbone of interchanging phosphorus and nitrogen atoms comprising two organic side groups supplementary to each phosphorus atom. Figure 4.13 shows the chemical formation of polyphosphazene, where R represents a variety of organic or organometallic side groups (Gunatillake and Adhikari, 2003; Nair and Laurencin, 2007).

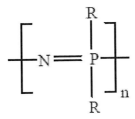

Figure 4.13. Chemical structure of polyphosphazene.

Polyphosphazenes that carry nonpolar amino acid esters are known to have excellent cell compatibility and near-neutral hydrolysis by-products. These polymers have been investigated yet again for applications ranging from stable compatible materials to control degrading materials for tissue engineering and drug delivery applications. Moreover, development of a class of biodegradable polyphosphazenes with hydrolytically penetrating organic margin groups, for instance amino acid esters, glucosyl, glyceryl, lactide and glycolide esters, have extend the awareness of biomaterial researchers (Nicolas et al., 2008; Lakshmi et al., 2003). It was reported that bioerosion of the polymer can be achieved by optimizing the hydrolysis rates of N-functional amino acid ester substituted polyphosphazenes, which decrease as the steric bulk of the substituent on the R-carbon of the amino acid residue increases (Chirila et al., 2002). Still, the synthesis of these polymers is simple as they only possess a single reactive site (the amino group) for intermingling to the polyphosphazene backbone. Sethuraman et al. 2006 stated that various multipurpose side groups could result in more to this class of polymers (Sethuraman et al. 2006).

4.3.2.12. *Polyphosphoester*

Polyphosphesters form different type of phosphorus encompassing polymers established as biomaterials. Moreover, polymers with phosphoester (P-O-C) replicatinglinkages in the backbone are of specific relevance in drug delivery research due to their biocompatibility and structural similarity to natural biomacromolecules like nucleic acids. The biodegradability of these polyphosphoesters is generated by hydrolysis or enzymatic scission of the ester bonds leading to harmless low molecular weight artefacts (Koseva et al., 2008).

Although routinely the polyphosphoesters can be structured by a various synthetic paths such as ring opening, poly condensation and polyaddition reactions, other techniques have also been reported by Nair and Laurecin (2007) and Lakshmi et al. (2003). The general structure of polyphosphoesters is shown in Figure 4.14, where R and R' can be varied to develop polymers with various physicochemical characteristics as in the case of polyphosphazenes.

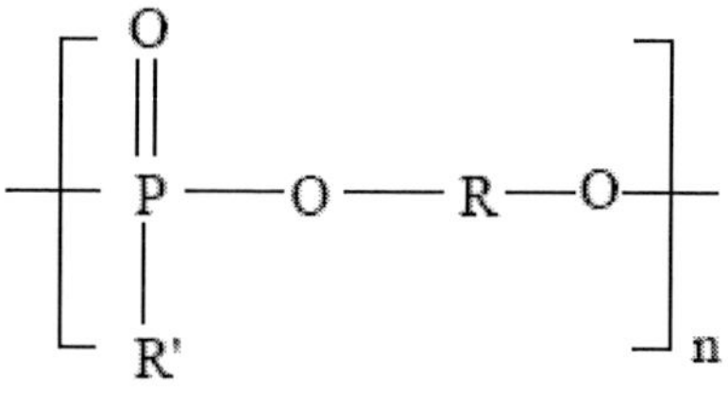

Figure 4.14. General structure of polyphosphoesters.

Moreover, i is recognised that the biocompatible poly(oxyethylene-H-Phosphonate)s affiliate with the polyphosphoester band are particularly attractive due to their excellent ability of controlling the rate of biodegradability together with the existence of highly reactive P-H group in developing components of the polymer chain (Koseva et al., 2008; Allcock et al., 2003). Penczek et al. (2005) and Zhao et al. (2003) also reported that polymers with duplicating phosphoester bonds in the backbone are physically adaptable. They are biodegradable on the account of hydrolysis, and under physiological conditions, enzymatic absorption at the phosphoester linkages could occur. These types of biodegradable polyphosphoesters are biocompatible and have the potential for a wide range of medical and pharmaceutical applications due to their similarity to bio-macromolecules such as nucleic acids (Penczek et al., 2005; Zhao et al., 2003).

4.3.2.13. *Hydrogel Materials*

These materials are cross-linked hydrophilic polymers, which can expand substantially when incorporated into a polar, solution. They have huge amount of water without dissolution, and they can keep their shapes with large percentage of water up to 99% by volume. This highly hydrated construction can be related to that of biological cartilages (Cheung et al., 2007). Moreover, hydrogels are readily used as injectable liquid to seal irregularly formed defects and widely used as catalysts to absorb with cells and bioactive agents. Hydrogels can be readily constructed to have appropriate elasticity to recover from the compressive deformation forces and continue in association with the loading part rather than the usage of fibrous mesh. Moreover, hydrogels can be applied to encapsulated cells and growth factors in polymer network (Cheung et al., 2007).

The summary of the main characteristics of synthetic materials are shown below in Table 4.3.

4.3.3. Properties of Materials for Soft Tissue Engineering

The physical properties of the selected materials used for manufacturing scaffolds for soft tissue engineering applications should include the following:

Tissue Compatibility: This implies that the scaffold material used should create minimum reaction with the tissue and should not cause any toxicity. Both of these properties of the scaffold material are very important regardless of the type of injury or healing on the internal surface. Generally, various tests are used to determine the biocompatibility of the scaffold material with that of the tissue being developed. For example, cell-cultured test is generally used *in vitro* to determine the tissue compatibility of the scaffold material.

Blood compatibility: Blood compatibility can be defined as the ability of the scaffold material as well as the blood to perform their functions effectively without causing any undesired reactions when both of them constantly interact with each other. This property of the material is of particular importance, especially for the biodegradable, bio-resolvable/ bio-absorbable polymers used as a scaffold material. This is primarily because these materials, when degraded or dissolved into the blood, should not prevent the blood from performing its functions. Ability of the material for its compatibility with the blood is determined with a series of *in vitro* and *in vivo* trials such as blood clotting tests (carried *in vitro*, e.g., the Lee-White test, platelet –adhesion test), and introduction of canine into the test site (carried *in vivo*).

Table 4.3. Summary of the main characteristics of synthetic materials

Polymer type	Investigators/developers	Main Characteristics	Drawbacks	Construction and Applications
Poly(glycolic acid) (PGA)	Jen et al., 1999, Mikos and Temenoff, 2000, Gunatillake and Adhikari, 2003, Shalaby and Burgh, 2004, Koseva et al., 2008	Due to its high crystallinity, it has excellent mechanical properties, and its degradation by-product glycolic acid is a natural metabolite.	Particular control of processing conditions is required as it has high sensitivity to hydrolytic degradation.	Solvent casting, particulate leaching can be used to fabricate porous foams—Applied as a bone internal fixation devices. Resorbable sutures.
Poly(lactic acid)(PLA)	Vert et al., 1984, William and Mort, 1997, Gunatillake and Adhikari, 2003, Lau, 2007, Cheung et al., 2007, Chu and Liu, 2008	Slower degradation rate than PGA. Excellent biocompatibility, Non-toxic and non-inflammatory. More resistant to hydrolytic attack than PGA.	Low tensile strength.	Fixation devices and musculoskeletal applications. In neural tissue engineering, applications PLA-PGA is used in bone repair applications.
Poly(caprolactone) (PCL)	In't Veld et al., 1997, Storey and Taylor, 1998, Gunatillake, Adhikari, 2003, Zhao et al., 2003, Dey et al., 2008	A biodegradable biopolymer. Superior mechanical properties. Low toxicity. Simple linear aliphatic polyester, strong and ductile.	Very low degradation rate Limits its use to controlled rate drug delivery. Deficiencies in cell binding domains.	PCL can be mixed with starch to lower its cost and, increase biodegradability. It can be added as a polymeric plasticizer to PVC. Long-term, implantable drug delivery system.
Poly (propylene fumarates)	Gunatillake and Adhikari, 2003, Wang et al., 2006, Efthimiou et al., 2011	Degrades by hydrolysis to fumaric acid and propylene glycol.	Has insignificant inflammatory response.	For the higher temperatures application, it could be used to further reinforce PPF-based materials. Used for drug formulations. Bone tissue engineering.
Polyanhydrides	Domb and Nudelman, 1995, Ibim et al., 1998, Muggli et al., 1999, Attawa et al., 1999, Gopferich and Tessmar 2002, Zhang et al., 2002, Vasheghani-Farahani and Khorram 2002, Gunatillake and Adhikari 2003, Timmer et al., 2003, Cui et al., 2004, Moffat and Marra 2004, Sarasam 2006, Nair and Laurencin 2007	Biodegradable, and biocompatibility. Excellent controlled release characteristics. It is cross-linkable. Non-flammable.	Requires storage under refrigeration, Low mechanical strength.	Injectable anhydrides could be used for filling irregularly shaped bone defects and soft tissue repair. Delivery of heparin for treating restenosis. Drug delivery system. Chemotherapeutic agents, Local anesthetics, anticoagulants, neuroactive drugs. Anticancer agents.

Polymer type	Investigators/developers	Main Characteristics	Drawbacks	Construction and Applications
Polyurethane	Wu et al., 2001, Gunatillake and Adhikari 2003, Baroli, 2006, Bonzani et al., 2007, Nair and Laurencin, 2007, Dey et al., 2008, Patel and Mequanint, 2009, Vlad et al., 2010,	Excellent biocompatibility and mechanical property. To improve cell adhesion, viability and proliferation. Ascorbic acid and glucose can be incorporated.	By using some diisocyanates, it creates some toxicity.	Various attempts have been made to formulate biodegradable polyurethanes. Medical applications include Long-term implants such as cardiac pacemakers Vascular grafts. Injectable hydrogels used for orthopaedic applications. Polyurethane based hydrogels for targeted drug delivery Tissue-engineering applications.
Poly (ether ester amide)	Lee et al., 2007, Nair and Laurencin, 2007, Patel and Mequanint, 2008,	Good mechanical and thermal properties. Hydrogen bonding ability of the amide bonds Biodegradability by ester bonds.	Difficult to fabricate due to the rise of emulsion formation	Attempts were made to increase the degradation rate of poly (ether ester amide) by adding amino acids. Medical applications include: Drugs and peptides
Poly(ortho esters)	Kumar et al., 2002 Nair and Laurencin, 2007, Paesen et al., 2007,	Desired mechanical and thermal properties. Hydrophobic and surface eroding polymer. Very slow erosion in aqueous environments The rate of degradation can be controlled by diols with varying levels of chain flexibility. Stable at room temperature. Useful pH sensitivity.	Poly(ortho esters) of class I had the lack of control over polymer erosion. Uncontrolled, autocatalytic hydrolysis reaction.	Good impact for Poly(ortho esters) of class III, on saturated drug delivery and eye though still under development. Specially used for drug delivery system. Ocular delivery, periodontal disease treatment, Tissue regeneration and applications in veterinary medical
Pseudopoly (amino acid)	Heller et al., 2002, Crisci et al., 2003, Nair and Laurencin, 2007,	Abundant, biocompatible and naturally occurring biodegradable polymers. .	Limited application due to immunogenicity and poor mechanical properties. Insolubility in common organic solvents, unpredictable water intake and swelling characteristics.	Under development for artificial skin substitutes, Drug delivery system, Bone conductivity

Table 4.3. (Continued)

Polymer type	Investigators/developers	Main Characteristics	Drawbacks	Construction and Applications
Poly (alkyl cyanocrylates)	Gupta and Lopina, 2002 Paesen et al., 2007, Nair and Laurencin, 2007, Vauthier et al., 2007,	Good biocompatibility. One of the fastest degradation polymers.	Could lead to toxic degradation for Alkyl side group.	Possible use as tissue adhesives. Possible use for gene delivery vehicles and oligodeoxynucleotides (ODN). Used for: Developing nano-particles, drug delivery system, skin applications.
Polyphosphazenes	Ratnar et al., 1996, Einmahl et al., 2001, Chirila et al., 2002, Crisci et al., 2003, Sethuraman et al., 2006, Nicolas et al., 2008,	Biodegradable, bioerodible material. Are hybrid polymers. High thermal stability.	For the solid-state pyrolysis of organo phosphazene/ organometallic (SSPO), it possesses long and difficult synthetic procedures.	Vaccine design. Tissue-engineering and drug delivery applications. Skeletal tissue regeneration. Used for rechargeable polymer lithium batteries, fuel cell membranes
Polyphosphoester	Lakshmi et al., 2003, Allcock et al., 2003, Zhao et al., 2003, Penczek et al., 2005, Diaz and Valenzuela, 2006, Nair and Laurecin, 2007	Biocompatible and symmetrical structure. Biodegradability increases by hydrolysis or enzymolysis scission of ester bonds.	Porosity control is difficult.	Effective gene carrier. Drug delivery applications. Possible for other biological and pharmaceuticals applications
Hydrogel Materials	Crisci et al., 2003, Cheung et al., 2007, Applications, 2011*	Acts as a facilitator to incorporate with cells and bioactive agents.	Silicon hydrogels, has decreased wettability, increased lipid interactions and accentuated lens binding.	Useful for cartilages and injectable scaffolds Contact lenses.

*http://research.chem.psu.edu/hragroup/applications.htm

Aging: is defined as rate of losing strength of the material after a certain period ot time. This property of the scaffold material is very important to ensure sufficient strength of the scaffold structure at all the time. Generally, any application of scaffold requires the scaffold to lose its strength with the course of the time, and this loss should be proportionate to the increase in the strength of the growing tissue so that the new tissue being developed does not fail during regeneration of the tissue and of sufficient strength to take up the functionality of the tissue or tissues being developed.

In addition to three above-mentioned properties, there are several other properties that are required in the material as a complete scaffold for tissue engineering:

- **Processability is the** ability to process complicated shapes with appropriate three-dimensional (3D) porous frameworks with the required degree of porosity which can reach over 90% in some cases. With such structure, high specific surface areas are required, which will benefit cell adhesions, generation and proliferation.
- **Workability and mechanical properties**. The matrix should have excellent workability and mechanical properties to support the new tissues generation.
- **Surface property between matrix and cells**. The matrix should provide good surface properties and good surface morphology to stimulate cell attachment and proliferation, and this is one of the most important factors in activating cell special gene expression and maintaining its phenotypic expression.
- **Demonstrate applicable** physical and biomechanical characteristics throughout the development of ECM regeneration activity.

However, it should be pointed out that depending on the application in hand, it is generally difficult to achieve all the above requirements and a compromise is always called for.

4.4. Final Remarks

Biodegradable polymers (BDP) are either derived from natural or synthetic sources. Process ability, modification and reproducibility are the domain of synthetic polymers. Natural occurring polymers are biocompatible, and they have highly controlled configurations and encompass an extracellular substance known as ligand, which can be bound to cell receptors. However, generally the availability is tricky as there are insufficient amounts of natural materials, and they are problematic in handling and moulding into scaffolds. Moreover, due to their superior properties in guiding the cells growth at different phases of progress, they may encourage an immune response at the matching time. In addition, their mechanical properties constrain their medical applications, and since the degradation of natural polymers is governed by the enzymatic processes, the rate of degradation can fluctuate from host to host and there is always the fear of antigenic and occurrence of diseases for allograft. However, generally, it is now well recognised that synthetic polymers in general pose superior physical properties in comparison with natural occurring materials (Cheung et al., 2007; Lau, 2007).

On the other hand, Synthetic polymers exemplify more constant source of raw materials, they are immunogenicity free and they can be constructed to produce broader range of physical properties to suite different tissue engineering applications. Different biodegradable polymers can be synthesized from the naturally occurring hydroxy acids, such as glycolic, lactic and ε-caproic acids with different cellular response in terms of degradation and reaction to the host (Balasundaram and Webster, 2007 and Khon J. and Langer, 1997).

At the present time, there have been a number of materials from natural origins or synthetic materials used to manufacture scaffold for clinical uses. As stated above, the materials used in tissue engineering field must have good biodegradability and biodegradable properties, which could be enzymolysis or animalcule hydrolysis with the existence of water. In general, an unstable backbone of the material leads to biodegradation. These materials usually contain chemical bond, which is easily hydrolysed, such as ester linkage and ether linkage. At present, polymers used in tissue-engineering applications are mainly polylactic acid (PLA), polyglycolic (PGA), polyhydroxybutyrate (PHB), polycaprolactone (PCL), polyanhydrides, polyphosphazenes, polyamino acid, pseudo-polyamino acid, poly(ortho esters), polyester urethane, polycarbonate, polyether, polyethylene glycol (PEG), polyethylene oxide (PEO), pluronic, polydioxanone, etc., and all these materials have good biocompatibility and do not express serious histologic reaction. In general, the parameters that influence the physical properties of biodegradable polymers include monomer and prime selections, the manufacturing settings, and the existence of additives. Moreover, one of the prime factors is the surface morphology, which should exhibit effective degree of porosity and should balance between the hydrophilicity and hydrophobicity for cellular attachment, proliferation and differentiation. Still, mechanical integrity of the scaffold should be of sufficient strength to match the surrounding tissue. The best regular chemical working groups with hydrolytically unstable linkages are esters, anhydrides, orthoesters and amides (Khon J. and Langer, 1997; Middleton and Arthur, 1998).

Concisely, presently used materials for tissue engineering applications of heart valve are not optimum but are good enough for the application in hand and offer several advantages such as similarity in properties and molecular structures to that of the natural tissue components. With the use of synthetic polymers, the physical properties can be altered and modified to meet the requirements for a particular application. With advancement in technology, new materials will be explored that would suit all the tissue-engineering applications and to suite different type of patients. Still, additive or active agents can be incorporated for fast tissue growth on the scaffolds. The ongoing research worldwide into the development of new DB and copolymers will no doubt lead to the development of various types of scaffolds for different TEHV-applications (Yang et al., 2001;Slaughter et al., 2009) It should be noted that here, we have only conducted a preliminary review of synthetic polymer variants, primarily because it is too hard to evaluate the claims made by different parties as to the benefits of their particular material characteristics.

References

Allcock, H. R., A. Singh, A. M. A. Ambrosio and W. R. Laredo, Tyrosine-bearing polyphosphazenes. *Biomacromolecules,* 2003. 4(6): pp. 1646–1653.

Applications, Available from: http://research.chem.psu.edu/hragroup/ applications.htm [Viewed on : 17 June, 2011].

Attawia, M.A., K. M. Herbert, K. E. Uhrich, R. Langer and C. T. Laurencin, Proliferation, morphology, and protein expression by osteoblasts cultured on poly(anhydride-co-imides). *J. Biomed. Mater. Res. (Appl. Biomater.)*, 1999. 48(3): pp. 322-327.

Balasundaram, G. and Webster, T.J., An overview of nano-polymers for orthopedic applications. *Macromol. Biosci.* 2007. 7(5): p. 635-642.

Baroli, B., Photopolymerization of biomaterials: issues and potentialities in drug delivery, tissue engineering, and cell encapsulation applications, *J. Chem. Technol. Biotechnol.*, 2006. 81(4): pp. 491-499.

Bonzani, I. C., R. Adhikari, S. Houshyar, R. Mayadunne, P. Gunatillake and M. M. Stevens, Synthesis of two-component injectable polyurethanes for bone tissue engineering, *Biomaterials*, 2007. 28(3): pp. 423-433.

Boccaccio, A, A. Ballini, C. Pappalettere, D. Tullo, S. Cantore, and A. Desiate, Finite element method (FEM), mechanobiology and biomimetic scaffolds in bone tissue engineering. *Int. J. Biol Sci.*, 2011. 7(1): pp. 112-32.

Basavaraju, K.C., T. Damappa, and S.K. Rai, Preparation of chitosan and its miscibility studies with gelatin using viscosity, ultrasonic and refractive index. *Carbohydr. Polym.*, 2006. 66(3): pp. 357-362.

Bai, X. P., H. X. Zheng., R. Fang, T. R. Wang, X. L. Hou, Y. Li, X. B. Chen and W. M. Tian, Fabrication of engineered heart tissue grafts from alginate/collagen barium composite microbeads. *Biomed. Mater.*, 2011. 6 (4): 045002.

Bencherif S. A., Synthesis, Characterization and Evaluation of Biodegradable Polymers, PhD Thesis, Carnegie Mellon University, 2009.

Crisci, L., C. D. Volpe, G. Maglio, G. Nese, R. Palumbo, G. P. Rachiero and M. C. Vignola, Hydrophilic Poly(ether-ester)s and Poly(ether-ester-amide)s Derived from Poly(ε-caprolactone) and -COCl Terminated PEG Macromers, *Macromol. Biosci.*, 2003. 3(12): pp. 749-757.

Chevallay, B. and D. Herbage, Collagen-based biomaterials as 3D scaffold for cell cultures: applications for tissue engineering and gene therapy. *Med Biol Eng Comput*, 2000. 38(2): p. 211-218.

Chirila, T. V., P. E. Rakoczy, K. L. Garrett, X. Lou and I. J. Constable, The use of synthetic polymers for delivery of therapeutic antisense oligodeoxynucleotides, *Biomaterials*, 2002. 23(2): pp. 321-342.

Cushion Joints with Hyaluronic Acid. Hyaluronic Acid News & Research; Available from: http://www.rejuvenation-science.com/hylauronic-acid_prof.html.—[Viewed on 16 June 2011].

Chu, P.K. and X. Liu, Biomaterials Fabrication and Processing Handbook, CRC Press, 7th April 2008.

Cui, J. F., Y. J. Yin, S. L. He, and K. D. Yao, Biodegradable polymeric scaffolds for bone tissue engineering. *Progress in Chemistry*, 2004. 16(2): pp. 299-307.

Cheung H. Y., K. T. Lau, T. P. Lu and D. H., A critical review on polymer-based bioengineered materials for scaffold development, *Composites Part B: Engineering*, 2007. 38(3): pp. 291-300.

Domb, A. J. and R. Nudelman, In vivo and in vitro elimination of aliphatic polyanhydrides. *Biomaterials*, 1995. 16(4): pp. 319-323.

Domb, A. J., N. Kumar and A. Ezra, Biodegradable Polymers in Clinical Use and Clinical Development, WILEY, 2011. Viewed from: http://books.google.com.au/books?on 16/11/11.

Diaz, C. and M. L. Valenzuela, Organomettalic Derivatives of Polyphosphazenes as Precursors for Metallic Nansostructured Materials, *J. Inorg. Organomet. Polym. Mater.,* 2006. 16(4): pp. 419-435.

Dey, J., H. Ju, J. Shen, P. Thevenot, S. R. Gondi, K. T. Nguyen, B. S. Sumerlin, L. Tang and J. Yang, Development of biodegradable cross-linked urethane-doped polyester elastomers, *Biomaterials,* 2008. 29(35): pp. 4637-4649.

Davachi S. M., B. Kaffashi, J. M. Roushandeh and B. Torabinejad, Investigating thermal degradation, crystallization and surface behaviour of L-lactide, glycolide and trimethylene carbonate terpolymers used for medical applications, *Materials Science and Engineering: C,* 2011. In Press, Corrected Proof.

Efthimiou, M., D. Symeonidis, G. Koukoulis, K. Tepetes, D. Zacharoulis and G.Tzovaras, Open inguinal hernia repair with the use of a polyglycolic acid-trimethylene carbonate absorbable mesh: a pilot study, *Hernia,* 2011. 15(2): pp. 181-184(4).

Einmahl, S., S. Capancioni, K. S. Abdellaoui, M. Moeller, F. C. Cohen and R. Gurney, Therapeutic applications of viscous and injectable poly(ortho esters), *Adv. Drug Delivery Rev.,* 2001. 53(1): pp. 45-73.

Features, Nonwoven and technical textiles, The Indian Textile Journal, 2011. Viewed from: http://www.indiantextilejournal.com/articles/FAdetails.asp?id=1242 on 23/11/11.

Gelse, K., Pöschl, E., Aigner, T., Collagens—Structure, function, and biosynthesis, *Advanced Drug Delivery Reviews,* 2003. 55 (12), pp. 1531-1546.

Gram, J., Five Exciting Industrial and Medical Uses of Collagen Article Source: http://EzineArticles.com/3075092.—[viewed on 15 June 2011].

Gunatillake, P.A. and R. Adhikari, Biodegradable synthetic polymers for tissue engineering. *Eur. Cells Mater.,* 2003. 5: pp. 1-16.

Gopferich, A. and J. Tessmar, Polyanhydride degradation and erosion. *Adv. Drug Delivery Rev.,* 2002. 54(7): pp. 911-931.

Gupta, A. S. and S. T. Lopina, L-Tyrosine-based backbone-modified poly(amino acids), *J. Biomater. Sci., Polym. Ed.,* 2002. 13(10): pp. 1093-1104.

Heller, J., J. Barr, S. Y. Ng, K. S. Abdellauoi and R. Gurni, Poly(ortho esters): synthesis, characterization, properties and uses, *Adv. Drug Delivery Rev.,* 2002. 54(7): p. 1015-1039.

HYALURONIC ACID. 2011; Available from: http://www.webmd.com/vitamins-supplements/ingredientmono-1062 viewed on 16 June 2011].

Hollinger J. O., Prelimianry report of osteogenic potential of a biodegradable copolymer of polylactide(PLA) and polyglycolide(PGA), *J. Biomed. Mater Res.,* 1983. 17: pp. 71-82.

Hargittai I. and M. Hargittai, Molecular Structure of hyaluronan: an intoduction, *Structural Chemistry,* 2008. 19(5): pp. 697-717.

Ibim, S.E.M., V. R. Sharti, K. E. Uhrich, S. F. El-Amin, R. Langer, R. Bronson and C. T. Laurencin, Preliminary in vivo report on the osteocompatibility of poly(anhydride-co-imides) evaluated in a tibial model. *J. Biomed. Mater. Res.,* 1998. 43(4): pp. 374-379.

Imeson, A., Food stabilisers, thickeners and gelling agents, John Wiley & Sons, 2006.

In'tVeld P.J.A., E. M. Velner, P. V. DeWhite, J. Hamhuis, P. J. Dijkstra and J. Feijen, Melt block copolymerisation of e-caprolactone and L-Lactide, *J. Polym. Sci. Part 1 Polymer Chem,* 1997. 35: pp. 219-226.

Jejurikar, A., G. Lawrie, D. Martin, and L. Grøndahl, A novel strategy for preparing mechanically robust ionically cross-linked alginate hydrogels. *Biomed. Mater.,* 2011. 6(2):025010.

Jayakumar R., D. Menon, K. Manzoor, S. V. Nair and H. Tamura, Biomedical applications of chitin and chitosan based nanomaterials- A short review, *Carbohydrate Polymers,* 2010. 82(2): pp. 227-232.

Jen A. C., S. J. Peter and A. G. Mikos, Preparation and use of porous poly (a-hydroxyester) scaffolds for bone tissue engineering. In: Tissue Engineering Methods and Protocols. Morgan J. R., M. L. Yarmush eds Humana Press, Totowa, 1999. pp: 133-140.

Koseva, N., A. Bogomilova, K. Atkova and K. Troev, New functional polyphosphoesters: Design and characterization, *React. Funct. Polym.,* 2008. 68(5): pp. 954-966.

Kakita, H. and H. Kamishima, Some properties of alginate gels derived from algal sodium alginate, in Nineteenth International Seaweed Symposium, Developments in Applied Phycology, M.A. Borowitzka, et al., Editors. *Springer Netherlands.* 2009. 2: pp. 93-99.

Knight, R. L., H. E. Wilcox, S. A. Korossis, J. Fisher and E. Ingham, The use of acellular matrices for the tissue engineering of cardiac valves. Proc. Inst. Mech. Eng., Part H: *Journal of Engineering in Medicine,* 2008 Jan., 222(1): pp. 129-143.

Kumar N., R. S. Langer and A. J. Domb, Polyanhydrides: an overview, *Adv. Drug Delivery Rev.,* 2002. 54(7): pp. 889-910.

King, J. H. Jr., Ocular collagen, human and animal: possible uses in ocular surgery. *Eye Ear Nose Throat Mon., Pub. Med.* 1969. 48(5): pp. 317-8.

Khon J. and Langer R., Bioresorbable and bioerodible materials. In: An Introduction to Materials in Medicine. Ratner B. D., Hoffman A. S., Schoen F. J., J. E. Lemon, eds. Academic Press, San Diego, 1997. pp: 65-73.

Kumirska J., M. X. Weinhold, M. Czerwicka, Z. Kaczynski, A. Bychowska, K. Brzozowski, J. Thoming and P. Stepnowski, Influence of the chemical structure and physiochemical properties of chitin and chitosan based materials on their biomedical activity, *Biomedical Engineering, Trends in Material Science,* edited by: Anthony L. Laskovski, Published by: In Tech. 2011.

Lau, K-T., Recent Development on Bio-composites Research, 16th International Conference on Composite Material, Kyoto, Japan, 2007.

Lee, K. W., S. Wang, B. C. Fox, E. L. Ritman, M. J. Yaszemski and L. Lu, Poly(propylene fumarate) Bone Tissue Engineering Scaffold Fabrication Using Stereolithography:_ *Effects of Resin Formulations and Laser Parameters, Biomacromolecules,* 2007. 8 (4): pp. 1077–1084.

Lakshmi, S., D.S. Katti and C.T. Laurencin, Biodegradable polyphosphazenes for drug delivery applications, *Adv. Drug Delivery Rev.,* 2003. 55(4): pp. 467-482.

Lee, C. H., A. Singla, and Y. Lee, Biomedical applications of collagen. *Int. J. Pharm.,* 2001. 221(1-2): pp. 1-22 .

Lodish H., A. Berk, S. L. Zipursky, P. Matsudaira, D. Baltimore and J. Darnell, Molecular Cell Biology, New York: W. H. Freeman; 2000. http://www.ncbi.nlm.nih.gov/books/NBK21582/ [Viewed on: 17/10/11].

Lahcini M., H. Qayouh, T. Yashiro, S. M. Weidner and H. R. Kricheldorf, Bismuth-Triflate-Catalized Polymerizations of ε-Caprolactone, Macromolecular Chemistry and Physics, 2011. 212(6): pp. 583-591.

Leng J. and A K. T. Lau, *Multifunctional Polymer Nanocomposites,* CRC Press, 2011.

Leadley S. R., K. M. Shakessheff, M. C. Davis, J. Heller, N. M. Franson, A. J. Paul, A. M. Brown and J. F. Watts, The use of SIMS, XPS and in situ AFM to probe the acid catalysed hydrolysis of poly (orthoesters), *Biomaterials,* 1998. 19(15): pp. 1353-60.

Mol, A., A. I. P. M. Smits., C. V. C. Bouten, and F. P. T. Baaijens, Tissue engineering of heart valves: advances and current challenges. *Expert Rev. Med. Devices,* 2009. 6(3): pp. 259-275.

Mendelson, K. and F. J. Schoen, Heart Valve Tissue Engineering: Concepts, Approaches, Progress, and Challenges. *Ann. Biomed. Eng.,* 2006. 34(12): pp. 1799–1819.

McNulty, M. Gelatin Available from: http://www.madehow.com/Volume-5/Gelatin.html.—[viewed on 15 June 2011].

McHugh, D.J., Production and utilization of products from commercial seaweeds, FAO Fisheries Technical Paper (FAO), no. 288 / Rome (Italy), FAO ,1987. (194 pages).

Moffat, K.L. and K. G. Marra, Biodegradable poly(ethylene glycol) hydrogels cross-linked with genipin for tissue engineering applications. *J. Biomed. Mater. Res., Part B- Appl. Biomater.,* 2004. 71B(1): pp. 181-187.

Muggli, D. S., A. K. Burkoth and K. S. Anseth, Cross-linked polyanhydrides for use inorthopedicapplications: Degradation behavior and mechanics, J. Biomed. Mater. Res., 1999.46(2): pp. 271-8.

Martina M. and D. W. Hutmacher, Biodegradable Polymers Applied in Tissue Engineering Research: *A Review, Polym Int.,* 2007. 56: pp: 145-157.

Middleton, J C. and Arthur J. T. Synthetic Biodegradable Polymers as Medical Devices. Originally published March 1998. Medical Plastics and Biomaterials Magazine, Medical Device Link.
http://www.devicelink.com/mpb/archive/98/03/002.html.

Mikos A. G. and J. S. Temenoff, Formation of highly porous biodegradable scaffolds for tissue engineering. *Electron J. Biotechnol,* 2000. 3: pp: 114-119.

Nicolas, J., F. Bensaid, D. Desmaele, M. Gronga, C. Detrembleur, K. Andrieux and P. Couvreur, Synthesis of Highly Functionalized Poly(alkyl cyanoacrylate) Nanoparticles by Means of Click Chemistry, *Macromolecules,* 2008. 41(22): pp. 8418–8428.

Nguyen, Thi-Hiep. and B.-T.Lee., Fabrication and characterization of cross-linked gelatin electro-spun nano-fibres. *J. Biomed. Sci. Eng.,* 2010. 3(12): pp. 1117-1124.

Nair L. S. and C. T. Laurencin, Polymers as Biomaterials for Tissue Engineering and Controlled Drug Delivery, *Adv. Biochem. Engin/Biotechnol,* 2006. 102: pp. 47-90.

Nair, L. S. and C. T. Laurencin, Biodegradable polymers as biomaterials, *Prog. Polym. Sci.,* 2007. 32(8-9): pp. 762-798.

Nelson J. F, H. G. Stanford and D. E. Cutright, Evaluation and comparison of biodegradable substances as osteogenic agents, *Oral Surg.,* 1977. 43: pp. 836-843.

Peter S.J., S. T. Miller, G. Zhu, A. W. Yasko and A. G. Mikos, In vivo degradation of a poly (propylene fumarate)/ ß-tricalcium phosphate injectable composite scaffold, *J. Biomed. Mater Res,* 1998. 41: pp. 1-7.

Petersen L. K., J. Oh, D. S. Sakaguchi, S. K. Mallapragada and B. Narashimhan, Amphiphilic Polyanhydrides Flims Promote Neural Stem Adhesion and Differentiation, *Tissue Engineering: Part A,* 2011. 17(19).

Paesen, B., T. Baekelandt, J. Heller, J. Martins and E. Schacht, Synthesis and complete NMR characterization of methacrylateendcapped poly(ortho-esters), *E-Polym.,* 2007.

Patil, K.P., D. K. Patil, B. L. Chaudhari, and S. B. Chincholkar, Production of hyaluronic acid from Streptococcus zooepidemicus MTCC 3523 and its wound healing activity. *Journal of Bioscience and Bioengineering,* 2011. 111(3): pp. 286-288.

Patel, A. and K. Mequanint, Syntheses and characterization of physically cross-linked hydrogels from dithiocarbamate-derived polyurethane macroiniferter, *J. Polym. Sci., Part A: Polym. Chem.,* 2008. 46(18):pp. 6272-6284.

Patel A. and K. Mequanint, Swelling kinetics of physically cross-linked polyurethane-block-polyacrylamide hydrogels, *Bioengineering, Procedings of the Northeast Conference, IEEE,* 2009.

Penczek, S, J. Pretula and K. Kaluzynski, Poly(alkylene phosphates): from synthetic models of biomacromolecules and biomembranes toward polymer-inorganic hybrids (mimicking biomineralization), *Biomacromolecules,* 2005. 6(2): pp. 547-51.

Pielichowska K. and Blazewicz S., Bioactive Polymer/Hydroxyapatite (Nano) composites for Bone Tissue Regeneration, *Advances in Polymer Science,* 2010, 232/2010, 97-207,

Qin, Y., Absorption Characteristics of Alginate Wound Dressings. *J. Appl. Polym. Sci.,* 2003. 91(2), pp. 953-957.

Rohanizadeh, R., M. V. Swain, and R. S. Mason, Gelatin sponges (Gelfoam®) as a scaffold for osteoblasts. *J. Mater. Sci. : Mater. Med.,* 2008. 19(3): pp. 1173-1182.

Research, Chitin, from
[http://www.ceoe.udel.edu/horseshoecrab/research/chitin.html]
on 11/11/11.

Ratnar, B. D., A. S. Hoffman, F. J. Schoen and J. E. Lemons, Biomaterials science: *an introduction to materials in medicine,* Elsevier Academic Press, 1996.

Rinaudo M., Chitin and chitosan: Properties and applications, *Prog. Polym. Sci.,* 2006. 31: pp. 603-632.

Rinaudo M., Behaviour of Amphiphilic Polysaccharides in Aqeous Medium, *TIP Rev. Esp. Cience. Quim. Biol.,* 2008. 11(1): pp. 35-40.

Rau S. W., *Development and Testing of a Machine-Coatable Chitosan Coating Applied to a Flexible Packaging Sealant, Thesis, Master of Science,* 2009.

Sabir M. I. , Xu X. and Li L., A review on biodegradable polymeric materials for bone tissue engineering applications, *Journal of Materials Science,* 2009, 44(21), 5713-5724.

Sarasam, A. R., Chitosan- Polycaprolactone Mixtures as Biomaterials—Influence of Surface Morphology on Cellular Activity, Submitted for PhD. to Oklahoma State University, 2006.

Schoen, F. J., Heart valve tissue engineering: quo vadis? *Current Opinion in Biotechnology,* 2011. 22(5). Schakenraad J. M., P. Nieuwenhues, I. Molenaar, J. Helder, P. J. Dykstra and J. Fiejen, In-vivo and in-vitro degradation of glycine/DL-lactic acid copolymers, *J. Biomed. Mater Res. 1989.* 23: pp. 1271-1288.

ScienceDaily, Math Model Could Aid Study Of Collagen Ailments, 2006, Available from:

http://www.sciencedaily.com/releases/2006/11/061114190020.htm.
- [viewed on 16 June 2011].

Shalaby, S. W., and K. J. L. Burg, Eds., Absorbable and Biodegradable Polymers—*Advances in Polymeric Biomaterials Series,* 2004: Taylor & Francis, Inc.

Shubhra, Q.T.H., A.K.M.M. Alam, and M.D.H. Beg, Mechanical and degradation characteristics of natural silk fibre reinforced gelatin composites. *Mater. Lett.,* 2011. 65(2): pp. 333-336.

Sethuraman, S., L. S. Nair, S. E. Amin, R. Farrar, M. T. N. Nguyen, A. Singh, H.R. Allcock, Y. E. Greish, P. W. Brown and C. T. Laurencin, In vivo biodegradability and biocompatibility evaluation of novel alanine ester based polyphosphazenes in a rat model, *J. Biomed. Mater. Res., Part A,* 2006.77A(4): pp. 679-687.

SWICOFIL, Chitosan, from
[http://www.swicofil.com/products/055chitosan.html] on 11/11/11.

Soares N. D. F. F., Chitosan-Properties and Application, Biodegradable polymer blends and composites from renewable resources, Edited by: Yu L., Published by John Wiley and Sons., 2009.

Slaughter, B.V., S.S. Khurshid, O.Z. Fisher, A. Khademhosseini, and N.A. Peppas, Hydrogels in regenerative medicine. *Advanced Materials*, 2009. 21(32-33): p. 3307-3329.

Storey R.F. and A. E. Taylor, Effect of stannous octoate on the composition, molecular weight, and molecular weight distribution of ethyleneglycol-initiated poly(e-caprolactone), *J. Macromol. Sci- Pure Appl. Chem.* 1998. A35: pp. 723-750.

Suprakas S. R. and Mosto B., Biodegradable polymers and their layered silicate nanocomposites: In greening the 21st century materials world, Progress in Materials Science, 2005, 50(8), 962-1079,

Timmer, M. D., R. A. Horch, C.G. Ambrose and A. G. Mikos, Effect of physiological temperature on the mechanical properties and network structure of biodegradable poly(propylene fumarate)-based networks. *J. Biomater. Sci., Polym.* Ed., 2003(A). 14: pp. 369-382.

Timmer M. D., C. G. Ambrose and A. G. Mikos, In vitro degradation of polymeric networks of poly(propylene fumarate)and the cross-linking macromer poly (propylene fumarte)-diacrylate, *Biomaterials,* 2003 (B). 24(4): pp. 571-577.

Tamura H., Recent research of biomedical aspectsof chitin and chitosan, Publishers: Research Signpost, 2008.

Ulery B. D., L. S Nair and C. T. Laurencin, Biomedical Applications of Biodegradable Polymers, Journal of Polymer Science Part B: *Polymer Science,* 2011.

Vauthier, C., D. Labarre and G. Ponchel, Design aspects of poly(alkylcyanoacrylate) nanoparticles for drug delivery, *J. Drug Targeting,* 2007. 15(10): pp. 641-663.

Vasilenko, A. P., N. V. Guzenko, and L. V. Nosach, Effects of modification of highly disperse silica with polyvinylpyrollidone on gelatin adsorption. *Pharm. Chem. J.,* 2010. 44(2): pp. 81-84.

Vázquez, J. A., M. I. Montemayor, J. Fraguas, and M. A. Murado, Hyaluronic acid production by Streptococcus zooepidemicus in marine by-products media from mussel processing wastewaters and tuna peptone viscera. *Microb. Cell. Fact.,* 2010. 9(46).

Vasheghani-Farahani, E. and M. Khorram, Hydrophilic drug release from bioerodible polyanhydride microspheres. *J. Appl. Polym. Sci.,* 2002. 83(7): pp. 1457-1464.

Vlad, S., C. Ciobanu, D. Macocinschi, D. Filip, M. Butnaru, L. M. Gradinaru and A. Nistor, Polyurethane nanofibres by electrospinning for biomedical applications, International Semiconductor Conference, 2010. pp: 353-356.

Vert M., P. Christel, F. Chabot, J. Leray, G. W. Hastings and P. Ducheyne, Macromolecular Biomaterials. CRC Press, Boca Ratan, 1984. pp: 119-142.

VanSliedregt A., A. M. Radder, K. deGroot and C. A. V. Blitterswijk, In-vitro biocompatibility testing of polylactides Part I: Proliferation of different cell types, *Mater Sci: Mater Med.*, 1992. 3: 365-370.

Verheyen C.C.P.M., J. R. deWijn JR, C. A.V. Blitterswijk, P. M. Rozing and K. deGroot, Examination of efferent lymph nodes after two years of transcortical implantation of poly (L-lactide) containing plugs: a case report. *J. Biomed. Mater Res.* 1993. 27: 1115-1118.

Wang, J. J., Z. W. Zheng, R. Z. Xiao, T. Xie, G. L. Zhou, X. R. Zhan and S. L. Wang, Recent advances of chitosan nanoparticles as drug carriers. *Int. J. Nanomed.*, 2011. 6: pp. 765–774.

Wang S., L. Lu and M. J. Yaszemski, Bone Tissue-Engineering Material Poly (Propylene fumarate): Correlation between Molecular Weight, Chain Dimensions, and Physical Properties, *Biomacrocolecules,* 2006. 7(6): pp. 1976-1982.

Willerth, S. M. and S. E. Sakiyama-Elbert, Combining stem cells and biomaterial scaffolds for constructing tissues and cell delivery, (July 09, 2008) StemBook, ed. *The Stem Cell Research Community, StemBook.*

William D. F. and E. Mort, Enzyme accelerated hydrolysisof polyglycolic acid. *J. Bioeng.,* 1977. 1. pp: 231-238.

Wu, L., H. Yu, J, Yan and B. You, Structure and composition of the surface of urethane/acrylic composite latex films, *Polym. Int.,* 2001. 50(12): pp. 1288-1293.

Yacoub, M. H. and J. J. Takkenberg, Will Heart Valve Tissue Engineering Change the World? *Nat. Clin. Pract. Cardiovasc. Med.,* 2005. 2(2): pp. 60-61.

Yang, S., Leong, K.F., Du, Z., and Chua, C.K., The design of scaffolds for use in tissue engineering. Part I. Traditional factors. *Tissue Eng.* 2001. 7(6): p. 679-689.

Zhang, T., M. Xu, H. Chen and X. Yu, Synthesis, degradation, and drug delivery of cycloaliphatic poly(ester anhydride)s. *J. Appl. Polym. Sci.,* 2002. 86(10): pp. 2509-2514.

Zhao, Z., J. Wang, H. Q. Mao and K. W. Leong, Polyphosphoesters in drug and gene delivery, *Adv. Drug Delivery Rev.,* 2003. 55(4): pp. 483-499.

Chapter V

Scaffold Fabrication Techniques

5.1. Introduction

As stated in the previous chapter, tissue-engineering scaffold is the first point of contact for development of the tissue on it, and, therefore, it should possess certain characteristics for proper growth of the tissue on it. The essential properties of the scaffold that are used in tissue engineering are its (1) Macro structure, (2) Pore size, Porosity and interconnectivity, (3) Surface area and surface chemistry, (4) Mechanical Properties, (5) Cell attachment, (6) Proliferation, (7) Differentiation and (8) Processing of variety of shapes and sizes (Hutmacher, 2001; Leong et al., 2003; Buckley and O'Kelly, 2004; Li et al., 2009; Sachlos and Czernuszka, 2003). These are some of the important properties for a good scaffold design for tissue-engineering applications. In addition to the above-mentioned characteristics, two other characteristics that can also be related to the material properties are that the scaffold surface should be acceptable for the growth of the tissue (i.e., it should be biocompatible with the tissue) and should be pharmacologically acceptable (i.e., it should be non-toxic, non-allergenic, etc.)

Moreover, there are several parameters that need to be controlled during the manufacturing process of a scaffold so as to ensure proper functionality of the scaffold. Each material used or combination of materials used for scaffolding possesses different processing requirements such as (1) Processing Conditions, (2) Process Accuracy, (3) Consistency, and (4) Repeatability. The different fabrication techniques are classified into conventional scaffold fabrication and Solid Free Form fabrications (Leong et al., 2003;Sachlos and Czernuszka, 2003). In this chapter, both techniques are discussed, including nanofabrication technique of electrospinning.

5.2. Scaffold Fabrication Methods

5.2.1. Conventional Techniques

The development of scaffold technology has been rapid, and it is worth putting the extent of technical advance in context by reviewing both the history and current fabrication techniques.

Conventional techniques of manufacturing scaffold make use of different biodegradable and bio-resorbable materials (Buckley and O'Kelly, 2004). These fabrication techniques involve (1) fibre bonding, (2) Gas foaming, (3) Emulsion freeze-drying, (4) Melt Moulding, (5) Membrane lamination, (6) Solvent casting and (7) Phase separation, which are commonly used for fabricating scaffold, and each has varying degree of success (Tan et al., 2003).

5.2.2. Gas Foaming

Many conventional scaffold fabrication techniques require organic solvent and high temperature for fabricating a scaffold, whereas for the gas foaming, there is no need for organic solvents or high temperature. The gas foaming process is widely used for PLGA polymer scaffolds, as the material and this type of fabricating scaffold combined to deliver high growth factors and cells (Ma and Elisseff, 2006). Gas foaming process takes advantage of the plasticizing properties of carbon dioxide and involves introducing the polymer in a high pressure CO_2 to obtain a saturated condition to get the carbon dioxide molecules dissolved into the polymers. When the gas pressure is decreased to ambient condition, a thermodynamic instability will occur; thereby, the carbon dioxide molecules dissolved in the polymer get unstable and separate from the polymers. The CO_2 molecules as they diffuse into the pore nuclei cause expansion of polymers and decrease in density. Thus, the resulting structure is the porous material, and the porosity obtained using this method is about 95% (Boland and Bowlin, 2008). Figure 5.1 below shows the principle of gas foaming process as described by Lim et al. (2008).

Another way of fabricating the scaffold is through combining both gas foaming and particulate leaching methods. In this technique, ammonium bicarbonate is added to a solution of methylene chloride, and the resulting solution is highly viscous, which can change the shape of the scaffold. The composite is then vacuum-dried after the solvent is removed. The purpose of vacuum-drying is to sublime the ammonium carbonate. This method produces porosity of 90%, with a pore size in the range of 200-500 micrometers. Figure 5.2 below shows the Gas foaming with Particulate Leaching process as described by Lim et al. (2008).

However, this technique also shows some limitations such as the usage of organic solvents, high temperature, and closed cell structure, as well as the fact that it is lacking pore interconnection and also time consuming (Duarte et al., 2009; Singh et al., 2010).

Scaffold Fabrication Techniques

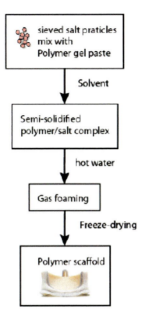

Figure 5.1. Ilustration of Gas Foaming Process.

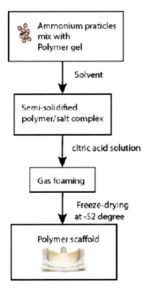

Figure 5.2. Illustration of Gas Foaming Process with particulate leaching.

5.2.3. Fibre Bonding

It is one of the conventional methods developed in 1990, to construct scaffolds for tissue engineering from a mesh of polymer fibres. This method is also called non-woven fabrics and can produce scaffold that has a large surface area to volume ratio. The fibre bonding technique consists of various processing methods, and it generally involves knitting or physical bonding of fibres produced by wet or dry spinning. Sometimes, electrospinning method will also be used to pre-fabricate the fibre. The main advantage of this scaffold fabrication method is that it has a large surface area for attaching the cell, and it promotes rapid diffusion of nutrients.

There are two methods of fabricating the scaffold using fibre bonding (Boland and Bowlin, 2008). Mikos and Temenoff (2000) developed a fibre bonding technique where the PGA fibres are embedded in the PLLA. These materials are then heated to above the melting point of both the polymers, which makes PLLA melt and fill the void present. Subsequently, the PLLA can be removed by dissolution using methylene chloride, resulting in a scaffold with highly porous material. The porosity obtained in this method is about 81% with a pore size of 500 µm. Another method of making the scaffold by fibre bonding is through atomization of PGA with PLLA or PLGA, which are used to coat the fibres. Here, chloroform is used to dissolve the PLLA or PLGA, and these dissolved materials are sprayed on to the PGA fibres. PGA has the property to be soluble in chloroform very slightly and because of that, the fibre remained unchanged during this process. The solvent is then removed, and PLLA or PLGA are glued to the fibres (Mikos and Temenoff, 2000). Although this process has many advantages, the porosity of the scaffold cannot be accurately controlled.

5.2.4. Phase Separation

The synthetic polymers are fabricated through thermally induced phase separation. In this method of phase separation, the polymer solution is separated into two phases by decreasing thermal energy of the polymer solution to induce a phase separation. The two phases obtained have one phase with high polymer concentration and the other with lower polymer concentration. With this process, the extraction, sublimation or evaporation process is used to remove the solvents that were present in the phase with low polymer concentration. This process makes open pores, and the polymer rich concentration gets solidified into skeleton of the polymer foam. It should be noted that the micro-and macro structure of the scaffold can be altered by changing the polymer concentration, temperature during phase separation and the types of polymer used for scaffolding.

The phase separation (Ma and Elisseff, 2006) can be of two different types such as liquid-liquid phase separation and solid-liquid phase separation. The liquid-liquid phase separation method generates polymer with poor and rich phases. The solid-liquid phase separation is where the temperature is low enough to allow the solution to freeze, which forms frozen solvent and concentrated polymer phases. This is also known as freeze-drying and is the most extensively employed method that could obtain a porosity of 90% or more. Note that the growth rates of ice crystals during the freeze-drying process determine the pore size of the scaffold (Ikada, 2006). Moreover, nano scale phase separation fabrication techniques can offer the following:

- By changing the operating conditions, the nature, dimension and density of the nanostructure scan can be optimized.
- High-density isolated Nano-dot-scan could be constructed in 3D structure (Zhang, 2011).

5.2.5. Particulate Leaching

Particulate leaching, also known as solvent casting, is widely used in scaffold fabrication for tissue-engineering applications. The porogen used in this method is Sodium chloride, due to its availability and ease of handling. In this method, the polymer is suspended in the solvent, and the salt particles are ground and sized into small particles and transferred into the mould. The suspended polymer is then casted into the salt-filled mould. The solvent is then removed by air or using vacuum, and the crystals formed are then leached away by immersing in the water, which forms a porous structure. The size of the porogen decides the pore size of the scaffold, whereas the salt and polymer composite ratio determines the porosity of the scaffold. Figure 5.3 shows illustration of the particulate leaching technique (Webster, 2007).

Technique of solvent casting particulate leaching uses a synthetic polymer added to the solvent solution along with salt. After the salt evaporates, porous scaffold of desired shape is left behind. This method of scaffold manufacturing suffers from following shortfalls:

- Improper connectivity of the pores inside,
- Difficulty in controlling the pores size,
- Evaporation rate of the solvent greatly influences the length of the pores-wall,
- Organic solvents from the scaffold must be carefully removed prior to its use for medical applications.

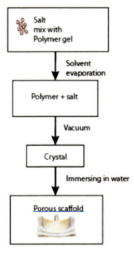

Figure 5.3. Particulate Leaching.

Although in general the solvent casting has many advantages, it still uses organic solvent, which is not favourable for tissue-engineering applications. Moreover, the pore size cannot be properly controlled, and the porogen used may have adverse effect on the biocompatibility of the scaffold.

The advantages and disadvantages of different conventional scaffold fabrication techniques as discussed by Leong et al. (2003), Morsi et al., (2008) and others are summarized in Table 5.1.

Table 5.1. Advantages and Disadvantages of Scaffold fabrication techniques

Conventional Method	Authors	% Attained Porosity	Advantages	Limitations	Achieved Pore size (µm)	Applications
Fibre Bonding	Boland and Bowlin, 2008, Mikos and Temenoff, 2000,	81	Interconnectivity among pores. It is very easy to process. It has high porosity and surface to volume ratio.	Harsh solvent, cell seeding not feasible. Non-amorphous polymer needs High processing temperature. Only limited polymers will be used. Lack of mechanical strength, control over the microstructure.	500	Tissue regeneration, atomization of PLLA or PLGA.
Solvent Casting/ Particulate Leaching	Mikos and Temenoff, 2000, Chuenjitkuntaworn et al., 2010	87	Good Mechanical properties of scaffolds could be achievd. It has high porous structure and has large pore size. Crystallinity can be controlled and, in turn, control porosity and pore size.	Harsh solvent, cell seeding not feasible. It is limited with membrane thickness The residual solvent and residual porogen are problems in this method.	100	Fabrication of PCL, PLLA and PLGA scaffolds, bone scaffolding application.
Gas foaming	Ma and Elisseff, 2006, Boland and Bowlin, 2008, Lim et al., 2008, Duarte et al., 2009, Singh et al., 2010, Montjovent et al., 2005, Montjovent et al., 2008	93	Good porosity. Plasticizing properties of carbon dioxide.	Lack of inter connectivity on surface. Usage of organic solvents, high temperature,, lacking pore interconnection and also time consuming. Cell seeding is difficult.	100	Cortical and trabecular bone repair and also human fetal bone cell, i.e., bone- tissue engineering.
Phase separation	Ma and Elisseff, 2006, Ikada, 2006, Zhang, 2011, Blaker et al., 2008	95	Good results in porosity and size of pores. It can produce the scaffold with high porous structure. Allows incorporating the agents which are bioactive.	Use of organic solvent prevents cell seeding. Lack of control over microstructure. It is limited to pore sizes and has problem with the residual solvents.	35	Localized drug delivery, tissue regeneration/augmentation and tissue engineering (for thermally induced phase separation, TIPS).

5.3. Textile Technique for Manufacturing Scaffold

As the nam implies this method is similar to the one that is used for producing fabric. In addition to tissue engineering of heart valve, non-woven type of textile technology has also been successfully used for tissue-engineering application of bone, cartilage, bladder and liver (Hutmacher, 2001). Non-woven type of scaffold manufacturing technique constructs scaffold by laying down long, continuous fibres of the scaffold material, and then these fibres are interconnected by means of chemical, mechanical, heat or solvent treatments. These types of scaffolds have a numbers of advantages over conventional types: have large surface area to volume ratio, and can be easily modified to optimize the physical and the mechanical properties to suit the application in hand (Edwards et al., 2004; Yarlagadda et al., 2005). However, this technique suffers from the drawback of lack of structural stability, and therefore the fibre bonding technique was developed to improve its mechanical properties to some extent.

5.3.1. The Principle of Membranes Lamination

The principle of membranes lamination is to manufacture a scaffold by laying down layer by layer sheets of the material to build up a complete scaffold. The sheets of the material are then laminated by using chloroform during the lamination process. This method of scaffold manufacturing suffers from the drawback of limited pore interconnectivity, use of highly toxic solvents and the enormous process time.

Though these techniques have been used from the past for manufacturing of scaffold, their shortcomings have initiated a need to develop a manufacturing process that would allow controlled manufacturing of scaffolds. Advanced manufacturing processes such as Solid Free Fabrication (SFF) and nanofabrication technology of electrospinning have proved to be the solutions for such kinds of problems. These are discussed below.

5.4 Rapid Prototyping

Rapid prototyping, also known as solid free fabrication (SFF), promises to become the most convenient and reliable technique for manufacturing of the scaffold in the coming years. Conventional scaffold manufacturing techniques used so far merely produced a scaffold that could be considered as a standard one (Tan et al., 2003). They produced a scaffold that served its purpose for functionality, but they have severe problems with control of the porosity and poor mechanical strength. Rapid prototyping, as the name indicates, was developed to produce prototype models of complicated and complex industrial designs so as to eliminate the cost incurred due to modifications required at the later stage. Prototype model instantaneously provides a replica of the model being produced by use of the design on the CAD system. It involves building the prototype model by layer-by-layer deposition of thermo material on the base material until the model is completed. Rapid prototyping produces components with a very high accuracy and precision and therefore can be considered a promising method for the manufacturing of the scaffold for tissue engineering. The most important advantage that the technique of rapid prototyping can offer is that it facilitates in the control of the porosity and interconnectivity of the scaffold by the use of the features available within the machine, thus making the scaffold much more functional to perform its role in a better manner (Yeong et al., 2004).

Rapid prototyping is available in different types based on the principle used to produce the object. Different types of commercially available rapid prototyping techniques used to manufacture scaffolds are discussed below:

5.4.1. Selective Laser Sintering (SLS)

This type of rapid prototyping technique uses a high power laser beam to fuse the powdered material to produce the required shape. The SLS equipment consists of a roller, a movable piston and a high-power laser beam (Figure 5.4).

In working, a roller is used to spread a layer of powder on to the base. The laser beam fuses the powdered material at the cross section of the object and produces a solid form on cooling. The unsintered powder is removed off, and a new layer of the powder is laid over it to trace and produce a new cross section of the object until the object is completed. This type of rapid prototyping technique has a resolution of about 500µm (Barnes and Harris, 2008) and is being clinically used to manufacture products made up of calcium phosphates, polymers such as PLA, PLGA, etc. (Chu and Liu, 2008).

To minimize powder wastage in constructing small prototyping, a compact SLS version is called for. Such a small machine may need additional electric supplies and different resourceful methods of sintering (Wiria et al., 2010).

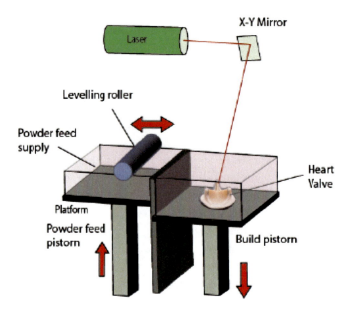

Figure 5.4. Schematic Illustration of Selective Laser Sintering.

5.4.2. 3D Ink-jet Printing

This method of rapid prototyping uses an additive to selective bond the loose powder dispersed in the shape of the cross section of the object defined on a CAD file (Figure 5.5).

In working, the polymer powder placed on the base material is fused together by spraying a liquid solvent through the nozzle. Subsequent layer of the material is then built up over the solid layer built below it by lowering down the table. This type of rapid prototyping system has a resolution of about 180µm (Barnes and Harris, 2008) and has been the choice for manufacturing of scaffolds made up of polymers such as PLA, PLGA, etc., and hydroxyapatite (Chu and Liu, 2008).

It should be pointed out that the development of the working materials is the main factor in establishing a technology for ultra-low-cost and large-area electronics. Moreover, ink-jet printing of metal NP ink is designed to utilize the huge melting temperature drop of the nanomaterial couple with the effortlessness of the NP ink invention. Such a low method temperature has considerable potential for the construction of electronics components on plastic substrates (Ko et al., 2010).

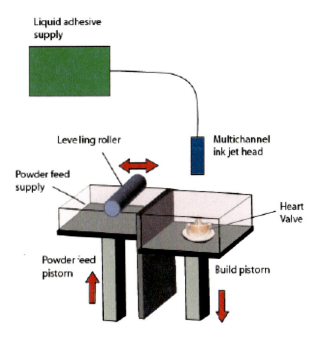

Figure 5.5. Schematic Illustration of 3D Ink-jet Printing.

5.4.3. Fused Deposition Modelling (FDM)

This method of rapid prototyping involves use of a heated nozzle to lay down the polymer in the visco-elastic form, which on solidification produces an object in accordance with the defined CAD geometry (Figure 5.6).

In working, material from the spool melts inside the liquefier head and then exits onto the built platform base through the nozzle. Nozzle assembly moves in the X and Y direction, whereas the platform moves in the Z-direction so as to provide 3D movement to produce 3D object in its complete form. This type of rapid prototyping technique has a precision of about 250μm (Barnes and Harris, 2008) and is being clinically used to produce scaffolds made up of bioerodible polymers (PLA, PLGA, etc.) (Shalaby and Burg, 2003).

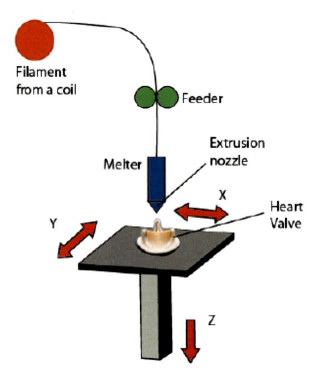

Figure 5.6. Schematic Illustration of Fused Deposition Modelling.

5.4.4. Stereolithography (SLA)

It is a liquid phase rapid prototyping technique that uses a UV light to solidify a photo-curable polymer liquid (Figure 5.7).

In working, energy from the laser beam is locally delivered to the photo-curable polymer liquid on the platform. The traced cross section is then lowered down, and a new subsequent layer is traced over it layer by layer to build up the complete object. This rapid prototyping technique has a resolution of about 250μm (Barnes and Harris, 2008) and is being clinically used to manufacture scaffolds only from the photo-polymer materials (Chu and Liu, 2008). .

The first kind of rapid prototyping process, namely Stereolithography (SLA), is widely used as it is known to have a better accuracy and degree of surface finishing with high level of smoothness and appearance than any other popular rapid prototyping processes such as FDM. There also exists a micro-version of stereolithography called "micro-stereolithography," which has the same working principle as the previous one. The only difference is that it uses a beam with smaller diameter so as to have a resolution of few microns

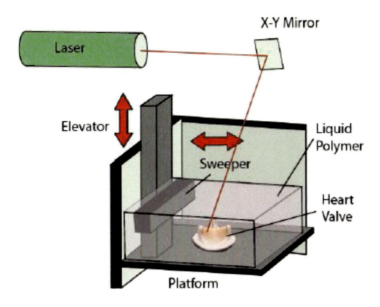

Figure 5.7. Schematic Illustration of Stereolithography.

However, like conventional techniques, Solid Free Fabrication (SFF) technique also suffers from some of the disadvantages. Table 5.2 summarizes the materials that are commonly used with a particular RP technique and limitations of each method as discussed initially by Chau and Liu (2008) and as discussed in this chapter.

5.5. Electrospinning for Manufacturing of a Scaffold

Electrospinning is a relatively old process that was developed over hundred years ago and has recently attracted attention for its use for manufacturing of scaffolds for tissue engineering, due to its ability to produce scaffolds of smaller diameter than those of the native cells. This property of the fibre helps proper organization of the cells around the fibre and thus encourages the growth of the tissue (Ramakrishna et al., 2005). The system is capable of accurately generating scaffold with optimized functionality and mechanical properties. Moreover, electrospinning can construct prototypes and scaffolds for various tissue engineering applications in nano scale with high degree of porosity and surface area.

Table 5.2. Rapid Prototyping Techniques and their limitations

Techniques	Investigators	Material used	RTM ratio (cm²/min)	Resolution μm	Advantages	Limitations	Applications
Membrane Lamination	Lanza et al., 2007	Bioerodible polymers (PLA, PLGA, etc.), bio-ceramics	Low<1	1000	The shape can be controlled and is independent of porosity and pore size.	Structures not really porous, low resolution. It lacks mechanical strength and has problem with residual solvent. It is a time-consuming process and has a limited interconnectivity between pores.	Tissue engineering.
3D printing	Barnes and Harris, 2008, Chu and Liu, 2008, Ko et al., 2010, Woesz 2008	Bioerodible polymers (PLA, PLGA, etc.), hydroxyapatite	Medium about 1	180	PLA, PGA, PLGA materials can be easily used	Presence of polymeric grains and excessive solvent. Limited resolution by nozzle size and particle size of the powder, Increased interconnectivity and porosity by removing unbound powder.	Production of metallic or ceramic scaffolds.
Laser Sintering	Barnes and Harris, 2008, Chu and Liu, 2008, Wiria et al., 2010, Woesz 2008	Calcium phosphates, polymers (PLA, PLGA, etc.)	Medium to high	<400	Good cosmetic appearance.	Presence of polymeric grains and excessive solvent. Less porosity, inter connectivity.	Fabrication of porous scaffolds made from polymers as well as fabricate structures (if high energy lasers and high heat resistance are used) from pure ceramic or metal powders, ideal for durable, functional parts with wide ranges of applications. Capable of producing snap fits and living hinges.

Techniques	Investigators	Material used	RTM ratio (cm2/min)	Resolution μm	Advantages	Limitations	Applications
Photopolymer isation/ Stereo Lithography	Barnes and Harris, 2008, Chu and Liu, 2008, Cooke et al., 2003	Photopolymeric resins	0.5 (medium)	250	PEG methacrylate, polyvinyl alcohol (PVA) and modified polysaccharides such as hyaluronic acid and dextran methacrylate can be used.	Use of photo-sensitive polymers and initiators, which may be toxic. Resolution is the biggest problem, post processing is required. Limited choice because of photopolymerizable material. Weak at the production time, post processing needed.	Bony substrates tissue engineering, excellent for fit and form testing, suitable for trade show-quality parts via painting and texturing.
Fused Deposition Modeling (FDM)	Barnes and Harris, 2008, Shalaby and Burg, 2003, Sachlos and Czernuszka, 2003, Muller et al., 2011	Bioerodible polymers (PLA, PLGA, etc.)	7 very high	250	PCL, And composite material can be used.	Limited to non-thermo-labile materials. Layered structure is very evident. Relatively bad surface finish, usage of relatively thick nozzle thus limits the resolution.	Bone tissue engineering, tissue-engineering scaffold production,
Techniques	Authors	Material used	RTM ratio (cm2/min)	Resolution μm	Advantages	Limits	Use in Relevant Fields
					Resulted in increase in cell seeding, variety of material can be used to manufacture various products.		Suitable for Conceptual models, Engineering models, and Functional testing prototypes. Good choice for high-heat applications, melting temperature is around 280°F.

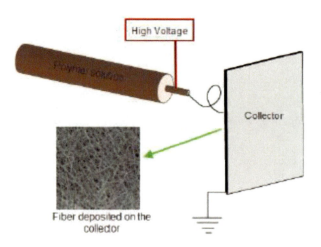

Figure 5.8. Schematic Illustration of Electrospinning.

Electrospinning equipment uses electrostatic forces for producing fibres from the polymer solution (Figure 5.8). Electrospinning equipment consists of a polymer solution placed in a tube-like vessel with a capillary tip to it. An electrode used to charge the polymer solution is placed in touch with the solution. A collector plate used to collect the ejected fibre from the capillary tube is placed in front of the capillary tube at a small distance apart and is connected to the ground that has a potential difference of about 10 to 30 kV. Due to the potential difference of the electric field, a charged jet is produced at the tip capillary tube known as Taylor Cone (Ramakrishna et al., 2005). The solvent from the jet evaporates as the jet travels towards the collector, leaving behind the charged fibre produced from the polymer. Collector plate being at lower potential attracts the charged fibre filament towards it and accommodates it on its surfaces after striking the plate. Between the time intervals from after the jet is ejected from the tube to the striking of the collector plate, the fibre initially retains a straight shape. However, later on, it becomes erratic due to the electric field acting on it. Thus, the process of electrospinning can be considered to consist of three steps, i.e., jet generation, evaporation of the solvent with fibre formation and solidification of the fibre on the collector plate (Ramakrishna et al., 2005). Electrospinning technique provides the advantage of control over the fibre dimensions and allows long and continuous fibres to be produced. Hence, it is a very cost-effective and a beneficial technique to produce nano-fibres as compared to the conventional techniques of producing nano-fibres such as phase separation, template synthesis, self-assembly, etc. Scaffolds made up form electrospinning have smaller diameter compared to that of the human tissue cells. It is believed that the human cells attach and organize well around the scaffold diameters of small size (Barnes and Harris, 2008). It is also easy to control the size and the length of the fibre being produced by controlling the processing parameters such as voltage, feed rate, temperature, type of collector, distance between the orifice and the collector (Ramakrishna et al., 2005).

However, this technique also poses some disadvantages such as usage of organic solvent is often required, jet instability and also some poor mechanical properties. Electrospinning is mainly used in the field of nano-fibres and also in tissue engineering, mesenchymal stem cell culture for bone and cartilage tissue engineering and so on (Lanza et al., 2007; Balta 2010;

Teo et al., 2011). However, the main problem that often occurs in electrospinning is beads formation. Beads are easily formed and decrease the surface area and deteriorate the function of the product (Huang et al., 2011).

5.5.1. Electrospinning Parameters

Electrospinning process is extremely dependent on various parameters. The reason for the electrospinning industrial process becoming complex is due to the complexity of process parameters. Based on the literature research, there are three main categories of parameters that affect the overall electrospinning procedure including:

- properties of the solution used as the feedstock (solution parameters),
- geometry and operation of the electrospinning apparatus (processing parameters),
- atmospheric and other local processing conditions (environmental parameters).

Thompson et al. (2007) have conducted an experiment with an electrospinning theoretical model with varying parameter values (Thompson et al., 2007). They reveal that there are five main factors affecting the process of electrospinning, such as volumetric charge density, distance from nozzle to jet, orifice radius, relaxation time and viscosity. In addition, solution parameters were initial concentration, solution density, electric potential, perturbation frequency, solvent vapour pressure. The next set of parameters includes environmental parameters such as relative humidity, surface tension, and vapour diffusivity. The first set of parameters affect the jet radius hugely and the second set moderately, and the final set affect the jet radius the least. The reason for using electrospinning is its long length, small diameters, and high surface area per unit volume. Therefore, the electrospinning operation has to be carefully monitored for the parameter variations. Most of the parameters affect the nozzle jet radius significantly. The smallest diameter occurs at the lowest flow rate of the solution. Fibre size and the production rate change with the solution concentration. The main objective of the experiment conducted by Thompson and the team was to investigate sensitivity of the theoretical model with the parameter variations. There are 13 effects with regard to its parameter variations are being analysed in the electrospinning model. All parameters were examined against the effect of the final jet radius. Some of the parameters analysed are mentioned below.

- The applied voltage affects the diameter at the end of the Taylor cone.
- Volumetric charge density is affecting the formation of the fiber beads for lower charge density solution.
- Decrease in fiber diameter in increasing the distance.
- Increasing viscosity gives a strong stretching of polymeric liquids.

Generally, based on the application in hand, researchers in the field can utilized these parameters and some other relevant parameters, to determine the most suitable model of the electrospinning process to use.

Nano-Fibrous Scaffolds for Tissue Engineering

These scaffolds are considered a huge breakthrough in bone and tissue engineering applications. Here a brief review of the use of this technique for selected applications is given.

As stated previously, in general, the main objective of tissue engineering is to develop scaffold to get a close resemblance of cellular matrix of the human organs. The design of the scaffold should have a proper pore size to be able to seed the cells and spread out through the matrix. When designing the scaffold, the main concern is to choose materials that can be biodegradable, biocompatible and non-toxic. Shalumon et al. (2011) have conducted a study for a suitable material for scaffolding. They have investigated using a blend of chitosan and polycaprolactone (CS/PCL) nano-fibres obtained by single step electrospinning. As stated in chapter 4, Chitosan is a widely used natural polysaccharide in many organic applications. Based on its ability to act as a biocompatible, biodegradable, antimicrobial active and wound healing material, it is most suitable for various tissue-engineering applications. Also, CS blended scaffolds are mechanically durable (Ohkawa et al., 2004; Geng et al., 2005; Chen et al., 2007). Moreover, according to Shalumon's study, this blend for the scaffold leads to a significant improvement in its hydrophobicity. As a result, the protein absorption and other bioactivity of the scaffold are being gradually increased. The characterization for hydrophobicity of the scaffold is done by measuring the contact angle with distilled water and Minimum Essential Medium (MEM). When the scaffold is mixed with PCL and CS, it gives a contact angle of 95.3±6 for water and 94.9±5 for MEM. Therefore, it is confirmed that this scaffold is hydrophobic.

Liao et al. (2011) conducted another study with electrospun Chitosan (CS). They fabricated multiwall carbon nanotubes (MWCNT) with electospun polyvinyl alcohol (PVA)/chitosan (CS) to obtain an improved cellular response for their scaffold and obtained the correct amount of water stability of the polymer. It was stated that one of the major important factors of using PVA is to obtain a good electrospinning ability to the CS. PVA is considered as a good water-soluble polymer. Mixing CS with PVA will make the blend with less intermolecular interaction of CS macromolecules (Duan et al., 2004).

Another widely used polymer blending is with poly(L-lactide)/poly (e-caprolactone)(PLLA/PCL). Liao et al. (2010) developed a porous structure with (PLLA/PCL) blend fibber membranes and a defect-free morphology and managed to obtain a high-performance fibre scaffold for various applications in tissue engineering. Moreover, they carried out a number of experiments to get the optimum diameter for electrospun fibre by varying the PLLA/PCL blend ratios. They reported that uniform 780nm diameter is considered as the optimum diameter of electrospun fibres for tissue engineering scaffolding.

As stated previously to acquire a good cell infiltration and a vascularization in tissue culture the scaffold should retain the correct microstructure and mechanical properties.. In order to obtain those properties , there has to be a good scale of pores on a scaffold. Sundararaghavan et al; (2010) have implemented a solution to improve the mechanical properties of a scaffold with a multi-scale porous structure. They used a technique called photo patterning to obtain a multi-scale porous structure. In this method, increasing the level of pores is achieved by a macromere solution containing a photo initiator that is exposed to light through a mask. Note that polymerization only occurs where light is transmitted. One of the key advantages in this photo patterning electrospinning method is that it forces the cells to travel in to dense structure by penetrating in to it.

5.6. Closing Remarks

There are two types of processes that are used for the manufacturing of tissue-engineered constructs—one is the conventional process and the second is SFF processes. The main purpose of using various techniques is to obtain high quality biocompatible porous scaffold. Conventional methods like, fibre bonding, solvent casting/particulate leaching, gas foaming/particulate leaching and liquid-liquid phase separation produce large, interconnected pores to facilitate cell seeding and migration. However, according to Mikos and Temenoff (2000), use of solvents in all of these methods is not attractive, as they are harsh and toxic and thus do not permit the cell seeding. Moreover, a long time is required for the removal of contamination, which does not allow feasibility in this regard as an immediate solution is to use them as a fabrication of scaffold.

Solid Freeform Fabrication is of prime importance to manufacturing scaffolds without defects and optimum porosity. 3D printing, Fused Deposition Modelling, Selective Laser Sintering, Stereolithography and Electrospinning are the devices that are explored for this purpose.

Stereolithography has limited viability in scaffold manufacturing. According to Wang et al. (1996), limited resolution and model shrinkage limit the use of 3D stereolithography in the scaffold manufacturing. Lee and Barlow (1993) were the first to use selective laser sintering process for the manufacturing of scaffold for bone implants.

Fused deposition modelling is one of the solid free-form techniques, which has been explored for the manufacturing of scaffold repeatedly and which holds promise in this regard as well. Woodfield et al. (2004) have produced scaffolds by using polyethylene glycol-terephthalate- Polybutylene terephthalate (PEGT/PBT) by the process of FDM. Good porosity and good mechanical strength and excellent geometry have been reported using this system. Chen and Morsi (Chen et al., 2010) have also explored the area of fused deposition modelling with electrospinning process to construct heart valve scaffold with aligned nano-fibres in leaflets, and thus an escalation in mechanical properties was reported by doing so. Thus, this hybrid technique is a key indicator for exploring the processes to obtain good mechanical properties and to attain effective porosity.

Concisely, use of SFF techniques for the manufacturing of scaffold can help us to develop scaffolds with uniform and controlled porosity, thus overcoming the problem of reproducibility of a scaffold or lack of it in others manufacturing techniques . Ongoing research with the use of nano fabrication technique of electrospinning for manufacturing of scaffold will help to produce scaffolds with structures similar to that of the tissue. There is no doubt that combining both techniques, SFF and Electrospinning, can yield an effective device for the manufacturing of scaffolds suitable for various tissue-engineering applications.

References

Balta, A. B., Development of Natural Compound-loaded Nanofibres By Electrospinning, M.SC. *Thesis submitted to İzmir Institute of Technology,* 2010.

Barnes, S. J. and L. P. Harris, *Tissue Engineering: Roles, Materials and Applications,* Nova Science Publishers, 2008.

Blaker, J. J., J. C. Knowles and R. M. Day, Novel fabrication techniques to produce microspheres by thermally induced phase separation for tissue engineering and drug delivery, *Acta Biomater.*, 2008. 4(2): pp. 264-272.

Boland, E. D. and G. L. Bowlin, Encyclopedia of Biomaterials ans Biomedical Engineering. Tissue Engineering Scaffold, Eds: Wnek, G. E. and Bowlin, G. L., *Informa Heathcare*, 2008.

Buckley, C. T and K. U. O'Kelly, Regular Scaffold Fabrication Techniques for Investigations in Tissue Engineering, Chapter V, *Topics in Bio-Mechanical Engineering*, TCD (Eds. P. J. Prendergast and P. E. McHugh), 2004, pp. 147-166.

Chen, R., C. Huang, Y. Morsi, X. M. Mo, Q. F. Ke and C. T. Donghua Univ, Fabrication Tissue Engineering Heart Valve Leaflet Scaffold Using Complex Electrospinning Method, International Forum on Biomedical Textile Materials, 2010. pp. 261-266.

Chen, Z., X. Mo and F. Qing, Electrospinning of collagen – chitosan complex, *Mater. Lett.*, 2007. 61(16): pp. 3490-3494.

Chu, P. K. and X. Liu, Biomaterials Fabrication and Processing Handbook, CRC Press, 2008.

Chuenjitkuntaworn, B., W. Inrung, D. Damrongsri, K. Mekaapiruk, P. Supaphol and P. Pavasant, Polycaprolactone/Hydroxyapatite composite scaffolds: Preparation, characterization, and in vitro and in vivobiological responses of human primary bone cells, *J. Biomed. Mater. Res., Part A*, 2010. 94A(1).

Cooke, M. N., J. P. Fisher, D. Dean, C. Rimnac and A. G. Mikos, Use of Stereolithography to Manufacture Critical-Sized 3D Biodegradable Scaffolds for Bone Ingrowth, *J. Biome. Mater. Res., Part B: Appl. Biomater.*, 2002. 64B(2).

Duan, B., C. Dong, X. Yuan and K. Yao, Electrospinning of chitosan solutions in acetic acid with poly(ethylene oxide), *J. Biomater. Sci., Polym. Ed.*, 2004. 15(6): pp. 797-811.

Duarte, A. R. C., J. F. Mano and R. L. Reis, Preparation of starch-based scaffolds for tissue engineering by supercritical immersion precipitation, *J. Supercrit. Fluid.*, 2009. 49(2): pp. 279-285.

Edwards, S. L., W. Mitchell, J. B. Matthews, E. Ingham and S. J. Russell, Design of Nonwoven Scaffold Structures for Tissue Engineering of the Anterior Cruciate Ligament, *Autex Research Journal*, 2004. 4(2).

Geng, X., O. H. Kwon and J. Jang, Electrospinning of chitosan dissolved in concentrated acetic acid solution, *Biomaterials*, 2005. 26(27): pp. 5427-5432.

Hutmacher, D. W., Scaffold design and fabrication technologies for engineering tissues—state of the art and future perspectives, *J. Biomater. Sci., Polym. Ed.*, 2001. 12(1): pp. 107-24.

Huang C., Y. Tan, X. Liu, A. Sutti, Q. Ke, X. Mo, X. Wang, Y. Morsi and T. Lin, 2011, Electrospinning of nanofibres with parallel line surface texture for improvement of nerve cell growth, *Soft Matter. 7*.

Ikada, Y., Ed., *Tissue Engineering: Fundamentals and Applications, Interface Science and Technology*, First Edition, Academic Press, 2006.

Ko, S. H., J. Chung, N. Hotz, K. H. Nam and C. P. Grigoropoulos, Metal nanoparticle direct ink-jet printing for low-temperature 3D micro-metal structure fabrication, *J. Micromech. Microeng.*, 2010. 20(12).

Lanza, R. P., R. S. Langer and J. Vacanti, *Principles of tissue engineering*, Burlington, MA: Elsevier Academic Press, 2007.

Lee G. and J. W. Barlow, Selective Laser Sintering of Bioceramic Materials for Implants, *Proceedings of Solid Freeform Fabrication Symposium,* 1993. pp. 376-380.

Leong, K. F., C. M. Cheah and C. K. Chua, Solid freeform fabrication of three-dimensional scaffolds for engineering replacement tissues and organs, *Biomaterials,* 2003. 24(13): pp- 2363-2378.

Li, M. G., X. Y. Tian and X. B. Chen, A brief review of dispensing-based rapid prototyping techniques in tissue scaffold fabrication: role of modeling on scaffold properties prediction, *Biofabrication,* 2009. 1(3).

LIAO, G. Y., L. CHEN, X. Y. ZENG, X. P. ZHOU, X. L. XIE, E. J. PENG, Z. Q. YE AND Y. W. MAI,Electrospun Poly(L-lactide)/Poly(e-caprolactone) Blend Fibres and Their Cellular Response to Adipose-Derived Stem Cells, *J. Appl. Polym. Sci.,* 2010. 120(4).

Liao, H., R. Qi, M. Shen, X. Cao, R. Guo, Y. Zhang and X. Shi, Improved cellular response on multiwalled carbon nanotube-incorporated electrospun polyvinyl alcohol/chitosan nanofibrous scaffolds, Colloids *Surf., B: Biointerfaces,* 2011. 84(2): pp. 528-535.

Lim, Y. M., H, J Gwon, J. Shin, J. P Jeun and Y. C. Nho, Preparation of Porous poly caprolactone scaffolds by gas foaming process and in vitro/In vivo degradation behavior using Gamma-Ray irradiation, *J. Ind. Eng. Chem.,* 2008. 14(4): pp. 436-441.

Ma, P. X. and J. Elisseeff, *Scaffolding in Tissue Engineering,* CRC Press, 2006. pp. 155-165.

Mikos, A. G. and J. S. Temenoff, Formation of highly porous biodegradable scaffolds for tissue engineering, *Electron. J. Biotechnol.,* 2000. 3(2), pp. 114-119.

Montjovent, M. O., S. Mark, L. Mathieu, C. Scaletta, A. Scherberich, C. Delabarde, P. Y. Zambelli, P. E. Bourban, L. A. Applegate and D. P. Pioletti, Human fetal bone cells associated with ceramic reinforced PLA scaffolds for tissue engineering, *Bone,* 2008. 42: pp. 554-564.

Montjovent, M. O., L. Mathieu, B. Hinz,, P. E. Bourban, P. Y. Zambelli, J. A. Mason and D. P. Pioletti, Biocompatibility of bioresorbable poly(L-lactic acid) composite scaffolds obtained by supercritical gas foaming with human fetal bone cells, *Tissue Eng.,* 2005. 11(11-12): pp. 1640-9.

Morsi, Y.S., Wong, C.S. and Patel, S.S., Conventional manufacturing process for three-dimensional scaffolds, *Virtual Prototyping of Biomanufacturing in Medical Applications,* 2008, Chapter7: pp. 129-148.

Muller, D., H. Chim, A. Bader, M. Whiteman and J. T. Schantz, *Vascular Guidance: Microstructural Scaffold Patterning for Inductive Neovascularization, Stem Cells International,* 2010 Dec., 2011, Article ID 547247.

Ohkawa, K., D. Cha, H. Kim, A. Nishida and Y. Yamamoto, Electrospinning of Chitosan, *Macromol. Rapid Comm.,* 2004. 25(18): pp. 1600-1605.

Ramakrishna, S., K. Fujihara, W. E. Teo, T. C. Lim and Z. Ma, An Introduction to Electrospinning and Nanofibres, 2005, from:http://www.worldscibooks.com /nanosci/5894.html [viewed on :16/06/2011].

Sachlos, E. and J.T. Czernuszka, Making Tissue Engineering Scaffolds Work. Review on the Application of Solid Freeform Fabricationtechnology to the Production of Tissue Engineering Scaffolds, *Eur. Cell. Mater.,* 2003. 5. pp. 29-40.

Shalaby, W. S, and K. J. L. Burg, *Absorbable and Biodegradable Polymers,* CRC Press, 2003.

Shalumon, K. T., K. H. Anulekha, K. P. Chennazhi, H. Tamura, S. V. Nair and R. Jayakumar, Fabrication of chitosan/poly(caprolactone) nanofibrous scaffold for bone and skin tissue engineering, *Int. J. Biol. Macromol.*, 2011. 48(4): pp. 571-576.

Singh, M., B. Sandhu, A. Scurto, C. Berkland and M. S. Detamore, Microsphere-based scaffolds for cartilage tissue engineering: Using subcritical CO_2 as a sintering agent, *Acta Biomater.*, 2010. 6(1): pp. 137-143.

Sundararaghavan, H. G., R. B. Metter and J. A. Burdick,Electrospun Fibrous Scaffolds with Multiscale and Photopatterned Porosity, *Macromol. Biosci.*, 2010. 10: pp. 265–270.

Tan, K. H., C. K. Chua, K. F. Leong, C. M. Cheah, P. Cheang, M. S. A. Bakar and S. W. Cha, Scaffold development using selective laser sintering of polyetheretherketone–hydroxyapatite biocomposite blends, *Biomaterials*, 2003. 24(18): pp. 3115-3123.

Teo, W. E., R. Inai and S. Ramakrishna, Technological advances in electrospinning of nanofibres, *Sci. Technol. Adv. Mater.*, 2011. 12(1), 013002.

Thompson, C. J., G. G. Chase, A. L. Yarin and D. H. Reneker, Effects of parameters on nanofibre diameter determined from electrospinning model, *Polymer*, 2007. 48(23): pp. 6913-6922.

Wang, W. L., C. M. Cheah, J. Y. H. Fu and L. Lu,Influence of process parameters on stereolithography part shrinkage, *Mater. Design*, 1996. 17(4): pp. 205-213.

Webster, T. J., Nanotechnology for the Regeneration of Hard and Soft Tissues, World Scientific Publishing Company, 2007.

Wiria, F. E., N. Sudarmadji, K. F. Leong, C. K. Chua, E. W. Chng and C. C. Chan, Selective laser sintering adaptation tools for cost effective fabrication of biomedical prototypes, *Rapid Prototyping J.*, 2010. 16(2): pp. 90-99.

Woesz, A., Rapid Prototyping to Produce POROUS SCAFFOLDS WITH CONTROLLED ARCHITECTURE for Possible use in Bone Tissue Engineering, Chapter 9, *Virtual Prototyping & Bio Manufacturing in Medical Applications, Springer*, 2008. pp. 171-206.

Woodfield, T. B. F., J. Malda, J. d. Wijn, F. Peters, J. Riesle C. A. v. Blitterswijk, Design of porous scaffolds for cartilage tissue engineering using a three-dimensional fibre-deposition technique, *Biomaterials*, 2004. 25(18): pp. 4149-4161.

Yarlagadda, P. KDV., M. Chandrasekharan and J. Y. M. Shyan, Recent Advances and Current Developments in Tissue Scaffolding, *Biomed. Mater. Eng.*, 2005. 15(3): pp. 159-177.

Yeong, W.Y., C.K. Chua, K.F. Leong, and M. Chandrasekaran, Rapid prototyping in tissue engineering: Challenges and potential. *Trends in Biotechnology*, 2004. 22(12): p. 643-652

Zhang, L., Self-assembled intrinsic nanoscale phase separation in polymers, *Europhys. Lett.*, 2011. 93(5).

Chapter VI

In Vitro Conditioning—
Bioreactors and ECM Generation

6.1. Introduction

In tissue engineering the scaffold made from either nature or polymer, of the correct configuration is seeded with one or more types of cells, and initially sited in culture medium (with required growth factors) and subsequently placed in controlled environment that simulates the physiological conditions in the human body. The biodegradable scaffold will interact with the surrounding tissues and degrades, leading to functional tissue structure (Morsi et al., 2007; Bilodeau and Mantovani, 2006).

In the literature, there are a number of researchers who reported successful implementations of various regenerated tissues (Hoerstrup et al., 2000; Martin and Vermette, 2005; Pörtner and Giese, 2006; Morsi et al., 2007). However, it can be stated that there has been a shortage in the delivery of tissue-engineering products, which is primarily due to the complexity and variation in the physiological conditions where the tissue is being established (*in vitro*) and where it is implemented for the function (*in vivo*). Regeneration of human tissues is very sensitive to any changes in the *in vitro* conditions, and consequently there is an urgent need to totally or partially mimic the *in vivo* conditions, which are hard to control during the regeneration process. In this process, the cells are influenced by the mechanical, electrical and chemical signals, which are essential for the cells' growth. The generation of these signals are indispensable for adequate cell differentiation, otherwise disorganisation or cell death can occur. This is because seeding cells on scaffold in static condition is not sufficient to regenerate the required 3D tissue. Living tissues need to acclimatize their physical structure and configuration to the living adjacent functional loads. Hence, it is now well recognised that *in vitro* conditioning via bioreactor is intensely important, at least in the initial stage, for adequate regeneration of the 3D tissue configurations (Galaction et al., 2007).

Hence, specially designed bioreactor for the application in hand is necessary for the initial development of tissue (Griffith and Naughton, 2002). Successively the concept of bioreactors has been introduced in tissue-engineering applications to control and create a reproducible environment and technical methods to simulate the physical conditions in clean and safe environments.

Although many investigations in tissue engineering have been conducted regarding manufacturing of scaffolds and seeding of cells onto scaffolds, less focus has been given to the design and manufacturing of bioreactors for tissue engineering in general, and tissue engineering of heart valve, in particular. In general, there are still a number of issues that need to be addressed to develop effective and efficient bioreactors (Martin and Vermette 2005; Morsi et al., 2007).

The organisational pattern of this chapter starts with the historical background of bioreactors, which gives an overview of when the concept of bioreactors was started and the different types of bioreactors evolved to date. Then the objectives of bioreactor will be discussed, followed by an outline of the working principles of different bioreactors and their capacities and limitations, with particular emphasis on bioreactor for heart valve conditioning.

6.2. Bioreactor for Tissue Engineering

As stated above, the basic purpose for the bioreactor is to ensure uniform distribution of tissue cells on the scaffold after seeding, to maintain adequate flow of the nutrients and supply gas through the scaffold, and to mimic the *in vivo* dynamic forces acting on the tissue. Moreover, it is now commonly accepted that the *in vitro* growth tissues in bioreactor must be accompanied and have relevant mechanical properties of the *in vivo* recipient's cells (Vismara et al., 2010). Therefore, those parameters such as oxygen mass transfer to the cells, waste mass transfer from the cells, and shear stress on cells to provide mechanical conditioning to mimic *in vivo* conditions need to be simulated (Morsi et al., 2007).

Berry et al. (2010) indicated that *in vitro* conditioning using specifically designed bioreactor can set to achieve two objectives. First, it simulates the biochemical and/or the hemodynamic settings required to regulate the interaction between the valvular cell types and surface material of the scaffold being conditioned. Secondly, it organises the entire cell-seeded valve scaffold before *in vivo* trials (Berry et al., 2010).

This chapter gives summary of most types of bioreactors available for tissue conditioning, including the ones used to condition TEHV prior to *in vivo* implantation.

6.2.1. Tissues Growth Factors

In designing a bioreactor system, the following factors should be considered so that an effective tissue generation can be maintained throughout the cell culture process. These are discussed below.

6.2.1.1. Waste Transfer

Waste transfer is as important a parameter as supply of nutrients to cells, but it is less significant when compared to oxygen supply to cells. Improper waste transfer can limit the final size of the tissues because of improper enzyme activities and toxicity due to either higher or lower level of waste accumulated. Therefore, within the design of a bioreactor provision, it should be ensured that there is proper waste transfer so that a healthy tissue is generated (Martin and Vermette, 2005).

6.2.1.2. Oxygen Transfer

Oxygen mass transfer is described as the relative movement of oxygen molecule to a fixed reference due to diffusion and convective flow. Generally, the relative movement of the oxygen should be sufficient to allow normal respiratory rate of the cells to avoid death of cells. Oxygen is the most important nutrient when compared to all the nutrients required for cells to grow into healthy tissue. However, oxygen transfer can, in some cases, limit the growth of the tissue when compared to other nutrients. The reason for this is the lack of good supply of sufficient amounts of oxygen to the surface of the cells because of less solubility of oxygen in the culture media. Therefore, to maintain good supply of oxygen to the cells, the medium should be re-oxygenated and circulated continuously via gas exchangers, and the exact amount of oxygen should be supplied, as either less or more amount of it could create adverse effects on the regeneration of tissues.

Moreover, different types of tissues need different levels of oxygen according to their construct. Though, enough oxygen should be supplied during the cell culture process to ensure that the cells grow into a healthy tissue. It is recognised that the required oxygen supply to the cells can be achieved by matching the diffusive transport with the convective effect (Martin and Vermette, 2005).

6.2.1.3. Hemodynamic Forces and Shear Stress

The bioreactor should be designed in such a way that a wide range of hemodynamic forces and shear stress can be generated.

The application of shear stress in the dynamic cell culture is another important factor in conditioning the cells to grow in required directions and provides required mechanical conditioning of the cells to sustain *in vivo* mechanical forces. It has been found by many researchers that shear stress has significant effect on endothelial cell proliferation by orienting the cell growth in the direction of fluid flow. Moreover, shear stress presents a determinant factor on tissue function and viability. Different tissues have different shear stress requirements for mechanical conditioning depending upon the maximum shear stress that they can sustain *in vivo.* Many studies have been conducted to obtain the required shear stress, but ideal shear stress and the required quantity and the hemodynamic conditions in bioreactors for various applications are yet to be determined (Martin and Vermette, 2005).

6.2.2. Types of Bioreactor for Cells Culture

In this section, a brief description of the most common bioreactors that have been used in tissue-engineering applications is given. There are basically two types of bioreactors for tissue regeneration, depending on whether the tissue needs to be developed in two- or three-dimensional (3D) structures. For two-dimensional, 2D, tissues such as skin for patients with skin burns, a simple culture vessel is observed by Martin and Vermette (Martin and Vermette, 2005). However, 3D tissues cannot be grown in simple culture vessels, as they require complex conditions to be implemented, such as proper supply of nutrients to the cells, transfer of waste materials from the cell and the correct shear stress and hemodynamic forces to be applied on the construct. Therefore, to develop 3D mammalian tissues, bioreactors for this purpose need to provide the essential environment with the ability of mass transfer and the correct shear stress magnitude and directions. Various bioreactors, which can grow 3D tissues, have been designed over past years for various applications. As noted above, different tissues require different nutrient supply, waste transfer and shear stress conditioning. Therefore, different bioreactors have been developed to provide the required conditions (Martin and Vermette, 2005; Morsi et al., 2007).

The designs of these bioreactors depend upon the type of the tissue to be developed. The bioreactor's size should be spacious enough to grow the tissue, and it should provide the required physical and chemical environment to grow the tissues and small enough to minimize the cost of the media. Galaction et al. (2007) classified the exciting bioreactors into two classes, rotating and non-rotating ones. With the first type, the rotating action can be easily controlled to produce a free-falling state that encourages the uniform growth of the tissues, to reduce the contact between the cells and the walls and to control shear stress to shield breakable tissues. Non-rotating bioreactor, on the other hand, is designed to generate specific biomechanical forces with a stationary culture chamber where the media can be designed to flow via the culture chamber and subsequently throughout the tissues (Galaction et al., 2007).

In the following sections, the different types of bioreactors for 2D and 3D tissue regenerations, depending upon the physical stimulation they provide to the cells *in vitro*, are briefly discussed.

6.3. Systems for Routine Cultivation

6.3.1. Petri Dishes

Petri dishes are first-generation bioreactors for organ culture (Figure 6.1). Although it is simple and efficient, the thickest tissues that are developed using Petri dishes reactors are approximately 0.5 mm thick tissue for a bone in an agitated dish under tension and 0.18 mm cardiac tissue in a static dish under tension. Therefore, Petri dishes reactors are not suitable for large sizes of tissues.

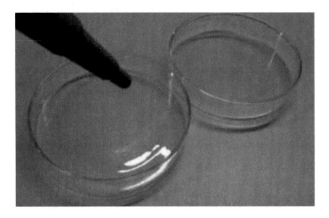

Figure 6.1. A Photo of Petri dishes.

6.3.2. Continuous Stirred-Tank Reactors

The continuous stirred-tank reactors are the first-generation bioreactors, which use static flasks and magnetically stirred flasks into which seeded scaffolds are suspended (as can be seen in Figure 6.2). These reactors thoroughly mix oxygen and nutrients in the culture medium and also reduce the boundary layer at the construct surface. These reactors stir at an average speed of 50-80 rpm. Cartilage tissues were grown in stirred reactors using both static and stirred reactor. However, the tissue grown in static reactor was found of less than 0.1 mm thickness and had a rough surface, whereas the tissue grown in stirred reactor was 0.3-0.5 mm thick and had a smooth surface. Moreover, even the thickness achieved for tissues (2-5 mm) in continuous stirred flask reactors are not suitable for clinically sized implants. The reasons for achieving tissues with very thin sizes are attributed to reactor's improper mass transfer to and from the cells. In another experiment, the seeded scaffolds are fixed, and turbulence is created by the supply medium around the constructs, which resulted in significant mass transfer, which assisted the growth of the tissue (Martin and Vermette, 2005; Galaction et al., 2007).

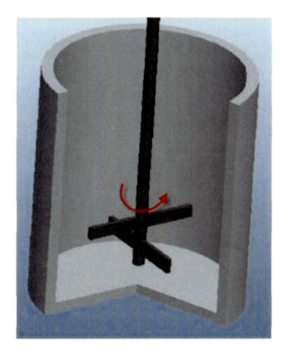

Figure 6.2. Pro/Engineer illustration of Stirred-Tank Reactors.

6.4. Systems for Continuous Cultivation

As shown in Figure 6.3, the required perfusion system for continuous cultivation consists of an oxygenator, a pump and a medium culture reservoir. With this design, the culture medium can be continually reintroduced with or without a supply of fresh medium to provide gas and nutrients to cells being conditioned and to eradicate metabolites and catabolites.

6.4.1. Hollow fibre reactors

This type of bioreactor was designed to provide more volume for cell culture and consists of bundles of hollow fibre within other fibres (Figure 6.4). These reactors are called coaxial hollow fibres and are specially developed to enhance transfer of nutrients and waste materials. It has been reported that a liver tissue has been developed using the coaxial bioreactor, which enhanced the mass transfer of nutrients and waste materials but limited the formation of tissue to the space inside the hollow fibres. Therefore, the tissue shape developed by this reactor is limited because it is formed in the shape of the tubes rather than uniform shape without any gaps, which limits their use for large tissue mass culture only (Mazzoleni et al., 2009).

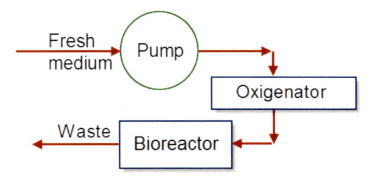

Figure 6.3. Schematic illustration of a simplified perfusion system.

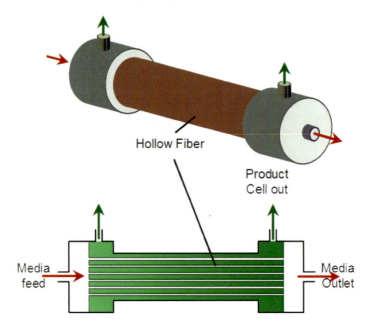

Figure 6.4. Schematic Illustration of the Hollow fibre reactor.

6.4.2. Rotating Wall Vessels

6.4.2.1. Rotating Perfused Wall Vessel

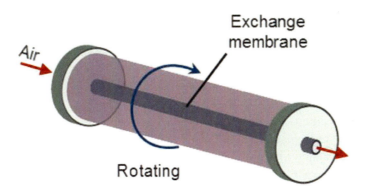

Figure 6.5. Schematic representation of Rotating perfused wall vessel.

Rotating Perfused Wall Vessel reactor, as initially discussed by Prewett et al. (1993), uses continuous renewal of the medium and applies swirling turbulent flow to it (Figure 6.5). This recirculation of the medium assists in controlling the amount of oxygen required, pH and temperature. These reactors are manufactured from two cylinders, one inside other, where the outer cylinder is a hollow one and the inner cylinder is a porous tube with a disc attached to it. Both cylinders, the inner with the disc and outer one, can be rotated independently at different speeds between 15 to 35 rpm. With this type of reactor, both cylinders can rotate as a solid-body at the same speed, and when the cells start accumulating, the speed of the rotation is increased to maintain uniform distribution of cell growth (Prewett et al., 1993).

However, to provide good transport of nutrients to cells and waste materials from cell, some fluid motion is required. With this microgravity process, a differential fluid motion is used, where the inner cylinder with the disk and outer cylinder are rotated at different speeds. The inner cylinder with the disk is rotated at higher speeds than that of the outer cylinder. This induces a secondary flow pattern in the radial-axial plane of the bioreactor, and as a result, the mixing of the medium is increased and mass transfer of the nutrients to the cells and waste materials from the cells are enhanced. To achieve good results, care should also be taken to avoid clustering of cells in one end or accumulation on the wall due to centrifugal force.

Still, with viscous pump bioreactor, the cells to form the tissues are delivered into the medium via small micro-carrier beads. These beads allow the cells to reproduce, and once enough cells are produced on each micro-carriers, the beads attach together to form into a large tissue. This viscous pump bioreactor is one of the well-known reactors for culture of mammalian cells and was developed at the University of Houston. Later, NASA developed it further to conduct experiments on earth and in space where experiments were conducted on shuttles STS-70 and STS-85, respectively, and also on the Russian space shuttle called MIR (Martin and Vermette, 2005).

6.4.2.2. High Aspect Ratio Vessel

High Aspect Ratio Vessel (HARV) is very similar to Rotating Wall Perfused Vessel (RWPV), and it was also developed by NASA. The difference between High Aspect Ratio Vessel and Rotating Wall Perfused Reactor is that the range of speed of HARV is lower in comparison to RWPV. Moreover, instead of maintaining speed in RWPV, the gas exchange in HARV is enhanced by using a larger gas exchange membrane at one end of the vessel. In fact, the area of the gas exchange of the HARV is double that of RWPV. Still various tissues, which are grown in RWPV, can also be grown in HARV (Martin and Vermette, 2005).

6.4.2.3. Perfused Chamber Bioreactors

With these types of bioreactors, the tissue constructs are fixed, instead of floating in the medium, while a continuous recirculation of the medium is carried out. This physical conditioning by the continuous flow of medium creates a good hydrodynamic shear force on the construct with cells, which in return encourages the growth of tissue with superior structure to sustain *in vivo* forces. The hydrodynamic forces that act on the construct provide good pressure fluctuations to stretch the cell membranes for good physical conditioning. These reactors are categorized into two types of reactors called perfused column and perfused chamber reactors. In perfused column reactors, structures are fixed onto the discs, which are stacked vertically, and the medium is circulated. In the other type of reactor, the construct is fixed onto the disc, which is placed horizontally on the base, and the medium is circulated (Pörtner et al., 2005).

6.5. Bioreactors for Tissue Engineering Heart Valves

Tissue engineering of heart valves requires the generation of the exact ECM that is capable of standing the correct physiological condition *in vivo*. This imposes a big challenge on the design of a bioreactor that is capable of totally simulating the required hemodynamic stresses and strains, and hence research and development of HV bioreactors is considered to be incomplete, as will be illustrated in this section.

Many researchers have created pulsatile flow of medium in the reactors and developed tissues like skin and cartilage (Martin and Vermette, 2005), but for culturing of cardiac tissues, a pulsatile medium condition through the centre of the construct is required. This necessitates the generation of pulsatile environment that totally or partially mimics the correct physiological condition of the heart. Hoerstrup created the earliest bioreactor design for tissue- engineered heart valves. They set out to create a "physiological flow and pressure for 'conditioning' the developing valve tissue construct prior to *in vivo* implantation" (Hoerstrup et al., 2000). This design utilised a pneumatic pump to provide flow through a silicone diaphragm, and it was first to use a pulsating flow for the cultivation of cells.

This proved to be advantageous to the construction of tissue-engineered heart valves as it showed that mechanical stresses are vital for the growth of valves that can survive the *in vivo* environment. Dumont et al. (2002) stated that the previous bioreactor designs such as Hoerstrup's showed insufficient mechanical integrity to withstand hemodynamic forces on the valve in the aortic position (Dumont et al., 2002). Due to this, Dumont created a new kind of pulsating bioreactor (Figure 6.6).

The outcome of this work produced a bioreactor that mimics three key components of the human heart—firstly, the left ventricle where the valve is grown, and secondly, the elastic aspect of the large arteries and finally, a resistance is applied to mimic the resistance caused by the arterioles and capillaries. To provide the mechanical flow of the fluid, a piston is used to change pressure and frequency throughout the differing stages of the culture growth. The **results showed that unlike previous designs, Dumont's bioreactor was able to provide both low- and high-pressure conditions to aid in extracellular matrix production.**

First developed in 2001, by Zeltinger, Zeltinger bioreactor utilises a closed loop circuit that is driven by a compressed air input to deform and expand a rubber bladder to create a pulsatile flow that passes through the tissue construct (Zeltinger et al., 2001). This basic design was followed by a similar one introduced by Layland et al. (Layland et al., 2003) (Figure 6.7), who altered the design by locating the valve at the bottom of the bladder rather than above.

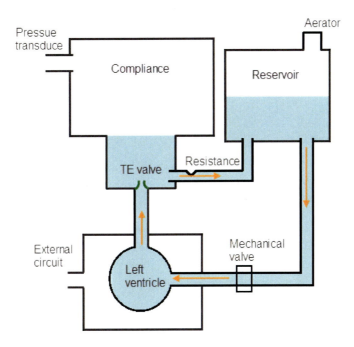

Figure 6.6. Schematic illustration of Prototype of Dumont's bioreactor (Dumont et al., 2002).

Tissue-engineered heart valves from Zeltinger's bioreactor showed cellular mass equivalent to 45% of a natural heart valve after eight weeks (Zeltinger et al., 2001). In comparison, the results of Layland group showed higher cellular mass than that seen in natural tissue. Also, the developed valve showed higher strength than a natural heart valve after 16 days of cultivation (Layland et al., 2003).

A new design concept was proposed by Engelmayr, Jr. et al. (2003) that used flexural simulation to provide tissue-engineered heart valves with a dynamic biomechanical environment to cultivate samples from. The bioreactor uses a linear actuator to provide uni-directional and bi-directional movement in the culture wells providing dynamic flexural stimulation. The simple movement of the flexural stimulation in Englemayr's design is capable of simulating the *in vivo* conditions as accurately as a pulse-duplicator (Engelmayr, Jr. et al., 2003). However, this design allows the isolation of flexure (from other potential modes of stimulation, such as pulsatile pressure or shear stress) (Engelmayr, Jr. et al., 2003) to allow for greater testing of the effects using flexural stimulation. In this experiment, the cultured cells showed greater strength and density than those cultured in static conditions.

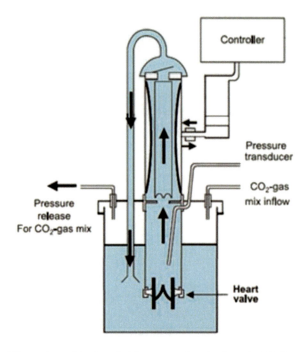

Figure 6.7. Schematic illustration of Layland's bioreactor (Layland et al., 2003).

Hildebrand et al. (2004) have developed a bioreactor that uses a stepper motor connected to an atrium, ventricle, compliance chamber and variable resistor (Figure 6.8).

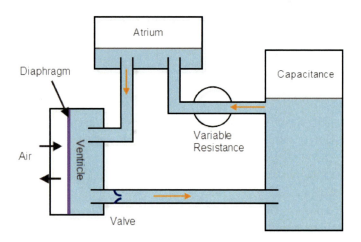

Figure 6.8. Schematic illustration of Bioreactor developed by Hildebrand et al., 2004.

This bioreactor design allowed for greater control of the differing parameters than previous designs. Adjustments can be made to the mean pressure, mean flow rate, beat frequency, stroke volume and the shape of the driving pressure waveform (Berry et al., 2010).

The bioreactor system designed by Mol et al. (2005) utilises diastolic pulse duplicator to create a pulsating flow (Figure 6.9). This is achieved by the use of two separate containers, one having the valve and one having the medium. A roller pump is used to compress air that contracts silicone tubes that push the medium into the chamber containing the valve, thus creating a pulsating flow.

The results from this experiment showed tissue culture over a four-week period that had undergone pre-strain or pre-strain combined with dynamic stress from the diastolic pulse duplicator. The tissue that formed showed a significant increase in both strength and stiffness over the tissue grown without the mechanical stimulation from the bioreactor.

A bioreactor developed by Lichtenberg et al. (2006) uses a closed circuit with a pulsatile pump, cell-seeding inlets on each side of the valve, compliance chamber and medium reservoirs to provide a continuous pulsatile flow to imitate pulmonary circulation (Figure 6.10).

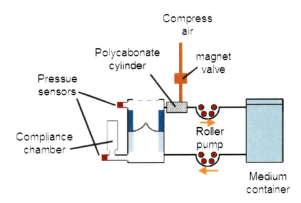

Figure 6.9. Schematic illustration of Bioreactor developed by Mol et al., 2005.

The heart valves **produced from this bioreactor "observed analogous biomechanical** properties when compared to native tissue" (Lichtenberg et al., 2006). The bioreactor also uses a varying flow that starts low and increases as the cultured tissue is formed. This bioreactor also showed increases in cell growth and adhesion.

Karim et al. (2006) developed a highly compact and simple bioreactor consisting of a small chamber connected to a pulsating pump and medium reservoir (Figure 6.11).

This bioreactor design operates in a simple closed-loop system. Although this bioreactor design does not have as many changeable parameters, the bioreactor still successfully achieved cellularization in the heart valve tissue.

Morsi et al. (2007) developed the bioreactor used at Swinburne University of Technology in 2007. It uses the same principle as that developed by Hoerstrup, with a diaphragm that pushes flow through the cultivating tissue via a compressed air input (Figure 6.12).

In Morsi's bioreactor, a pneumatic pulse duplicator is utilized to provide the input for the system, and this allows for a wide range of pressures and waveforms to be produced. Due to this, the design can be used across a variety of applications including tissue-engineered heart valves.

The bioreactor design created by Syedain and Tranquillo (2009) uses a syringe pump to provide cyclic pressure and stretching of valve located in the latex tube. A peristaltic pump is also used to provide a constant flow of the culture medium (Figure 6.13).

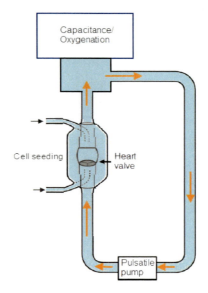

Figure 6.10. Schematic illustration Bioreactor developed by (Lichtenberg et al., 2006).

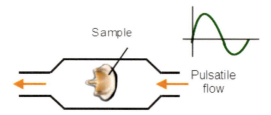

Figure 6.11. Schematic of Karim's Bioreactor growth chamber (Karim et al., 2006).

Figure 6.12. Bioreactor used at Swinburne University of Technology (Morsi et al., 2007).

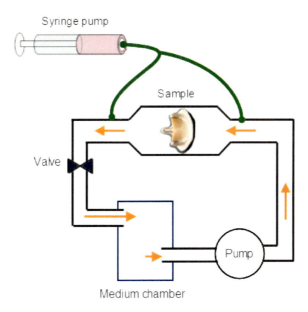

Figure 6.13. **Schematic illustration of the Syedain & Tranquillo's Bioreactor.**

This bioreactor design is able to increase the pressure and amount of stretch applied on the valve as the tissue is cultured. The results showed large improvements when compared with statically grown valves; however, all showed similar thickness. The valves also showed similar radial attributes to natural heart valves, in tensile stress, maximum tension and membrane stiffness. However, when tested radically, the engineered valves showed lower values for tensile stress and maximum tension, and comparable value for membrane stiffness. These results show greater findings of the variable aspects of the valves, where previous studies have not.

In order to exhibit independent and coupled stimulatory effects of TEHV, in 2008, a Flex-Stretch-Flow (FSF) bioreactor was designed by Engelmayr Jr. et al. (2008). The FSF bioreactor consisted of two identical chambers, which could hold up to 12 rectangular tissue **specimens (25 x7.5 x1 mm) via a novel "spiral-bound" tissue**-gripping technique, and the flow was generated by magnetically coupled paddlewheel. The specimens can be changed from stretch to curvature by controlling the moving aim. This bioreactor provided a robust tool for the study of mechanical stimuli on *in vitro* engineered HV tissue formation and allowed studying the effects of spatially varying shear stresses.

Schleicher et al. (2010) demonstrated a real-time pulse bioreactor for tissue engineering under various physiological and haemodynamic conditions. The scaffold of the heart valve was placed in a transferring plate in culture medium within a cylindrical culture reservoir made from medical grade materials for placement in a 37 ^{0}C incubator (Figure 6.14). The pulse rates and pressures inside of the culture reservoir were adjusted and controlled by a microprocessor-controlled system. This design simulates the correct physiological conditions and allows continuous monitoring of them during the *in vitro* conditioning. This system is easy to operate and allows up to eight heart valves to be tested simultaneously .

In another study of bioreactor (Durst and Allen, 2010) an effective bioreactor system capable of conditioning various type of biological tissues has been proposed. In describing the **bioreactor it was stated that "the system consisted of pistons, which were** driven in the culture medium cylindrical chambers by a crank and cam assembly (Figure 6.15). The stroke volume (651 mL) and frequency (0–110 BPM) of the flow through the sample could be controlled by varying the crank length and shaft speed of the motor". Moreover, it was highlied that one of the main limitations of the design was the lack of good wave form control and feedback system.In addition, this system requires large volumes of media and an entire half-height of standard incubator to operate (Durst and Allen, 2010).

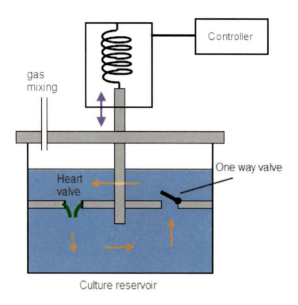

Figure 6.14. Schematic illustration of Simplified pulse reactor (Schleicher et al., 2010).

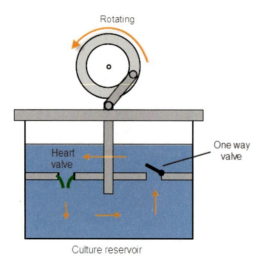

Figure 6.15. Schematic illustration of Synchronous multivalve aortic valve culture system (Durst and Allen, 2010).

In 2010, Vismara et al. developed an effective controlling system within the bioreactor to control the mechanical forces in real time (Figure 6.16) (Vismara et al., 2010). This bioreactor was described as a laboratory-oriented tool, which could carry out *in vitro* culture trials under strictly controlled conditions. The controlling the compliance allows the determination of the cells being conditioned in real time. In the screening phase, the compliance index could be useful in the overall monitoring of the process (Vismara et al., 2010).

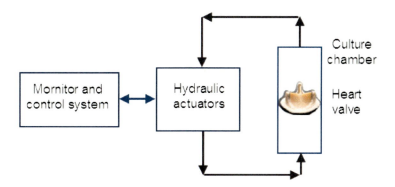

Figure 6.16. Schematic of compliance monitoring bioreactor for TEHV (Vismara et al., 2010).

Moreover, Bowles et al. (2010) constructed a pulsatile bioreactor, controlled by stepper motor and computer interface (Figure 6.17). This bioreactor consists of piston cylinder arrangement, attached to the outer chamber of a pulsatile sac ventricular assist device with one way mechanical valve. The adjustable piston created the correct flow circulation of tissue culture medium under various operating conditions. This bioreactor offers a high degree of control and reproducibility of hydrodynamic operating conditions, where pressure and flow can be independently adjusted. These features may prove advantageous in the development of a conditioning regime, which produces a tissue-engineered valve with effective features for human implementations.

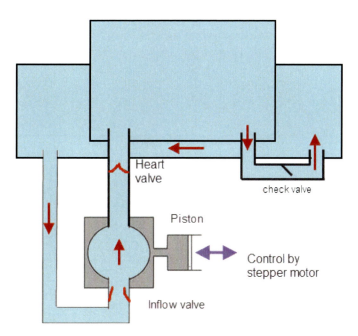

Figure 6.17. Schematic illustration of the Bowles et al. computer-controlled stepper motor bioreactor.

Table 6.1. Summary of Bioreactor researches

Authors	Main uses	Type of mechanical stimulation	Advantages	Disadvantages
Hoerstrup et al., 2000	Tissue-engineered heart valves	Pneumatic pump moving silicone diaphragm.	-Pulsating flow with backflow	-Limited variation of mechanical stimulation
Dumont et al., 2002	Tissue-engineered heart valves	Piston driven flow.	-Pulsating flow with backflow - Variation of pressure	-Design has limited available variations for experimentations
Zeltinger et al., 2001,Layland et al., 2003	Tissue-engineered heart valves	Compressed air/vacuum expands and contracts rubber bladder, creating a flow through growing valve.	-Closed circuit decreases risk of contamination	-Complex design in nature -Single sample application
Engelmayr et al., 2003	Tissue-engineered heart valves	Linear actuator providing bilateral movement in culture wells.	-Large amount of samples cultured at a time -Simple design -reduced risk of contamination in each culture well	-Only bilateral stimulation applied to samples
Hildebrand et al., 2004	Tissue-engineered heart valves	Stepper motor in closed circuit.	-Closed circuit reduces risk of contamination	-Not capable of providing backflow
Mol et al., 2005	Various strain-based applications	Roller pump using compressed air to contract and expand silicone tubes controlled via a diastolic pulse duplicator.	-Various stress and strains able to be applied -Various pulses are able to be replicated -Multiple samples able to be cultured simultaneously	-Only linear mechanical stimulation can be provided to the samples via flow passing through
Lichtenberg et al., 2006	Tissue-engineered heart valves	Closed circuit with pulsatile Pump.	-Closed circuit reduces risk of contamination	-Limited amount of mechanical stresses that can be applied
Karim et al., 2006	Tissue-engineered heart valves	Closed circuit with pulsatile Pump.	-Closed circuit reduces risk of contamination -Highly compact design	-Limited amount of mechanical stresses that can be applied
Morsi et al., 2007	Various tissue culture	Pneumatic pump controlled by pulse duplicator moving silicone diaphragm.	-Can be used for various applications -Various signals can be replicated by the pulse duplicator	-Limited amount of mechanical stresses that can be applied
Engelmayr et al., 2008	Tissue-engineered heart valves	Flow was generated by magnetically coupled paddlewheel.	-Allows study of mechanical stimuli on *in vitro* engineered HV tissue formation and allowed study the effects of spatially varying shear stresses	-Limited by the shape of the investigated specimens

Authors	Main uses	Type of mechanical stimulation	Advantages	Disadvantages
Syedain & Tranquillo, 2009	Tissue-engineered heart valves	Syringe pump controlling the compression and expansion of latex tube that has constant medium flow.	-High amount of variation of stress and strains can be experimented with.	-Pulsating flow not to be replicated
Schleicher et al., 2010	Tissue-engineered heart valves	Pulse rates and pressures inside of the culture reservoir were adjusted and controlled by a microprocessor-controlled system.	-Easier to operate, reduces the risk of contamination, and up to eight HV can be studied simultaneously	
Durst and Allen, 2010	Organ culture and tissue engineering	Controlled by varying the crank length and shaft speed of the motor.	-Easier to operate, reduces the risk of contamination	-Lack of robust flow control and feedback systems and imprecise waveform control
Vismara et al., 2010	Tissue-engineered heart valves	Hydraulic actuator.	-Real-time, non-destructive, and non-invasive monitoring	
Bowles et al., 2010	Organ culture and tissue-engineered heart valves	Computer –controlled stepper motor.	-Offers a high degree of control and reproducibility of hydrodynamic operating conditions	

6.6. Final Remarks

All the bioreactors discussed above are either using hydrodynamic or mechanical forces or both to create continuous physical conditioning for cell cultures. Pulsatile bioreactors are devices that can provide mimicking *in vivo* environment of tissue culture by simulating the correct physiological, mechanical and hydrodynamic forces. Applying such conditions on the cells during the culture ensures that the tissues developed possess the required properties of the natural tissues.

Bioreactors play an important role in tissue engineering by providing the required environment for the cells seeded on the scaffold to grow and differentiate into healthy tissues, which can replace damaged organs in human body and perform the assigned function and sustain the *in vivo* forces. When the concept of bioreactors originally started, it took place as simple bioreactors called as Petri dishes and spinner flasks, which only provided the required culture medium. These bioreactors did not provide any mechanical and hemodynamic conditioning to make the scaffold able to sustain *in vivo* forces. The next generation of bioreactors included the rotating bioreactors such as Slow-Turning Lateral Vessel (STLV), Rotating Wall Perfused Vessel (RWPV) and High-Aspect-Ratio Vessel (HARV), which provided proper mixing of the medium to provide good mass transfer to the construct. There were two kinds of rotations for these bioreactors. One is called solid rotation, where both the cylinders are rotated at same speed in the same direction, such that the constructs spin together with the walls; the other type is called free-fall rotation of microgravity environment. This environment provided tissues with better cell differentiation and uniform distribution of cells. New designs of these reactors are capable of providing the correct physiological conditions, sufficient amount of oxygen and nutrient supply, temperature, pH, and transport of catabolites and metabolites. From this chapter, it can be stated that rotating bioreactors are more suitable for conditioning and culture of delicate tissues; for complex dynamic constructs such as heart valve, non-rotating bioreactors are widely utilized and adapted (see Table 6.1).

To date, these above-mentioned bioreactor designs have improved significantly since the introduction of the first concept of bioreactor and truly allow well-differentiated, three-dimensional tissues with specific physical properties (Bilodeau and Mantovani, 2006). The next generation of bioreactors include the perfusion bioreactors, which used the process of sending the culture medium through the constructs to create stresses on the construct and provide it with better mechanical conditioning to sustain *in vivo* forces. Then, the concept of imitating the *in vivo* environment was developed by the pulsatile flow bioreactors. Pulsatile flow bioreactors provide in general the cyclic flow of medium through the constructs by using respiratory pumps to create the pulsatile flow. These bioreactors in general provide better mechanical conditioning of the tissues and uniform differentiation of the cells.

Researchers have started to develop bioreactors that can culture various types of tissues instead of just sticking to one tissue. Narita and colleagues developed a bioreactor that is capable of producing physiological pressures and able to interchange between heart valves and blood vessels, but the design is complicated because it consists of four separate interconnected chambers. Morsi et al. (2007), on the other hand, developed a bioreactor that is able to produce pulsatile flow related to cardiac cycle, which imitates the systolic and diastolic phases. This Pulsatile flow developed is described as physiological pressures *in vitro* and condition, required for cells such as smooth cells and endothelial cells to make them able to sustain hemodynamic forces *in vivo*. The designed bioreactor is described as able to culture both arteries of different length and diameters and heart valves of different sizes by changing the adapters attached to the perfusion chamber and using proper perfusion chambers. This bioreactor developed has three interconnected chambers, whereas the bioreactor developed by Narita and colleagues has four interconnected chambers.

It should be noted that many of the articles relating to bioreactor design are not specific to heart valves (e.g., may relate to bones), but the information is still relevant. Even though the work of Pathi et al. (2005) is about using a bioreactor for the growth of other matters, the ideas found in the article are still valid for growing heart valves. The article highlights that the amount of nutrients the cells in the bioreactor receive has an impact on the growth of the cells and how the design of bioreactor can greatly influence the amount of nutrients received.

The work of Lim et al. (2009) is about designing a bioreactor with the optimum flow for optimum growth of cells. Again, this is not specifically about designing a bioreactor for heart valves, but its ideas and the theory are the same and can be applied to the design of bioreactors for other applications. This article suggests that large increases in growth rate can be achieved by optimising the bioreactor. Campolo et al. (2009) look at the flows and the surfaces of the bioreactor and use numerical simulations to help modify the design of a bioreactor to create flow conditions that will maximise cell growth. The most recent article on bioreactor design is by Osborne et al. (2010), who closely investigate the relationship between the design and mechanical environment of the bioreactor and how it affects the growth and quality of tissue it produces. This research uses computer-based finite element methods to test how the geometry of the bioreactor influences the flows and the internal environment.

To sum up, the generation of biomechanical stresses and strains are essential for the differentiation, but they should have the right magnitudes and directions to suite the cultivated tissues. For example, the kinds of loading and mechanical stresses required for cartilage culture are certainty not suitable for cardiac tissues. Moreover, the correct pulsatile physiological hemodynamic of stresses and strains are essential during the culture phase of blood vessels and heart valves to generate the correct structure of ECM and increase their consistency and their mechanical integrity and functionality. However, in any case, oxygenation is beneficial in all *in vitro* conditioning to increase cellular absorption in a culture medium. An additional important point in multifarious tissue formation, which is not covered in this chapter, is the regeneration of an adequate vascularization for efficient function of tissue integration with the host (Bilodeau and Mantovani, 2006). Nevertheless, the general consensus among researchers that whether the constructs grown in the pulsatile and other bioreactors satisfied, up to some level, the ideal *in vitro* conditions required to grow the tissue in a bioreactor is not known yet. Research is being carried out to establish the ideal conditions for *in vitro* development of the vascular constructs by applying different strategies to create the ideal flow of the medium.

References

Berry, J. L., J. A. Steen, J. K. Williams, J. E. Jordan, A. Atala and J. J. Yoo, Bioreactors for development of tissue engineered heart valves, *Ann. Biomed. Eng.,* 2010. 38(11): pp. 3272-3279.

Bilodeau, K. and D. Mantovani, Bioreactors for Tissue Engineering: Focus on Mechanical Constraints. *A Comparative Review, Tissue Eng.,* 2006. 12(8): pp. 2367-2383.

Bowles, C. T., S. E. P. New, R. V. Loon, S. A. Dreger, G. Biglino, C. Chan, K. H. Parker, A. H. Chester, M. H. Yacoub and P.M. Taylor, *Hydrodynamic Evaluation of a Bioreactor for Tissue Engineering Heart Valves, Cardiovascular Engineering and Technology,* 2010. 1(1): pp. 10-17.

Campolo, M., F. Beux and A. Soldati,Transport of species in a bioreactor for bone tissue engineering, Icheap-9: *9th International Conference on Chemical and Process Engineering,* Pts 1-3, 2009. 17: pp. 1179-1184.

Dumont, K., J. Ypeman, E. Verbeken, P. Segers, B. Meuris, S. Vandenberghe, W. Flameng and P. R. Verdonck, Design of a New Pulsatile Bioreactor for Tissue-engineered Aortic Heart Valve Formation, *Artif. Organs.,* 2002. 26(8): pp. 710-714.

Durst, C. A. and K. J. G. Allen, Design and physical characterization of a synchronous multivalve aortic valve culture system, *Ann. Biomed. Eng.,* 2010. 38(2): pp. 319-325.

Engelmayr, G. C., Jr., D. K. Hildebrand, F. W. H. Sutherland, J. E. Mayer, Jr. and M. S. Sacks, A novel bioreactor for the dynamic flexural stimulation of tissue engineered heart valve biomaterials, *Biomaterials,* 2003. 24(14): pp. 2523-2532.

Engelmayr Jr, G.C., Rabkin, E., Sutherland, F.W.H., Schoen, F.J., Mayer Jr, J.E., and Sacks, M.S., The independent role of cyclic flexure in the early in vitro development of an engineered heart valve tissue. *Biomaterials,* 2005. 26(2): p. 175-187

Engelmayr ,G. C. Jr., L. Soletti, S. C. Viqmostad, S. G. Budilarto, W. J. Fiderspieal, K. B. Chandran, D. A. Vorp and M. S. Sacks, A novel flex-stretch-flow bioreactor for the study of engineered heart valve tissue mechanobiology, *Ann. Biomed. Eng.,* 2008. 36(5): pp. 700-712.

Gupta V. and K. J. Grande-Allen, Effects of static and cyclic loading in regular extracellular matrix synthesis by cardiovascular cells, *Cardiovascular Research,* 2006. 72: pp: 375-383.

Galaction, A. I., D. Cascaval and E. Folescu, Bioreactors for 3D tissue engineering, 2007. [Viewed from: http://ebooks.unibuc.ro/biologie/RBL/vol12nr6/1.htm] on 23/06/11.

Griffith, L. G. and G. Naughton, Tissue Engineering—Current Challenges and Expanding Opportunities, *Science,* 2002. 295(5557): pp. 1009-1014.

Hildebrand, D. K., Z. J. Wu, J. E. Mayer, Jr. and M. S. Sacks, Design and hydrodynamic evaluation of a novel pulsatile bioreactor for biologically active heart valves, *Ann. Biomed. Eng.,* 2004. 32(8): pp. 1039-49.

Hoerstrup, S. P., R. Sodian, J. S. Sperling, J. P. Vacanti and J. E. Mayer, Jr., New pulsatile bioreactor for in vitro formation of tissue engineered heart valves, *Tissue Eng.,* 2000. 6(1): pp. 75-79.

Hoerstrup, S. P., A. Kander, S. Melnitchouk, A. Trojan, K. Eid, J. Tracy, R. Sodian, J. F. Visjager, S.A. Kolb, J. Grunenfelder, G. Zund and M. I. Turina, Tissue Engineering of

Functional Trileaflet Heart Valves From Human Marrow Stromal Cells, *Circulation,* 2002. 106(12 Suppl 1): pp. 1143-50.

Karim, N., K. Golz and A. Bader, The cardiovascular tissue-reactor: a novel device for the engineering of heart valves, *Artif. Organs,* 2006. 30(10): pp. 809-814.

Kelly D. J. and P. J. Prendergast, Mechano-regulation of stem cell differentiation and tissue regereration in osteochondral defects, *J. Biomech,* 2005. 38(7): pp: 1413-22.

Lichtenberg, A., I. Tudorache, S. Cebotari, S. R. Lichtenberg, G. Sturz, K. Hoeffler, C. Hurscheler, G. Brandes, A. Hilfiker and A. Haverich, *In vitro re-endothelialization of detergent decellularized heart valves under simulated physiological dynamic conditions, JOURNAL ¿?*2006. 27(23): pp. 4221-4229.

Lim, K. T., P. H. Choung, J. H. Kim, H. M. Son, H. Seonwoo, and J. H Chung,Development of a perfusion bioreactor system for alveolar bone tissue engineering , Published by: *American Society of Agricultural and Biological Engineers,* 2009.

Layland, S. K., F. Optiz, M. Gross, C. Doring, K. J. Halbhuber, F. Schirrmeister, T. Wahlers and U. A. Stock, Complete dynamic repopulation of decellularized heart valves by application of defined physical signals—an in vitro study, *Cardiovasc. Res.,* 2003. 6D(3): pp. 497-509.

Martin, Y. and P. Vermette, Bioreactors for tissue mass culture: Design, characterization, and recent advances, *Biomaterials,* 2005. 26(35): pp. 7481-7503.

Mazzoleni, G., D. D. Lorenzo and N. Steimberg, Modelling tissues in 3D: the next future of pharmaco-toxicology and food research?, *Genes Nutr.,* 2009. 4(1): pp. 13-22.

Mol, A., N. J. Driessen, M. C. Rutten, S. P. Hoerstrup, C. V. Bouten and F. P. Baaijens, *Tissue Engineering of Human Heart Valve Leaflets: A Novel Bioreactor for a Strain-Based Conditioning Approach, Ann. Biomed. Eng.,* 2005. 33(12): pp. 1778-88.

Morsi, Y. S., Birchall I. E. and F. L. Rosenfeldt, Artificial aortic valves: an overview, *Int. J. Artif. Organs.,* 2004. 27(6): pp. 445-51.

Morsi, Y. S., W. W. Yang, A. Owida and C. S. Wong, Development of a novel pulsatile bioreactor for tissue culture, *Int. J. Artif. Organs.,* 2007. 10(2): pp. 109-114.

Osborne, J. M., R. D. O'Dea, J. P. Whiteley, H. M. Byrne and S. L. Waters, The influence of bioreactor geometry and the mechanical environment on engineered tissues, *J. Biomech. Eng.,* 2010. 132(5).

Pathi, P., T. Ma and B. R. Locke, Role of nutrient supply on cell growth in bioreactor design for tissue engineering of hematopoietic cells, *Biotechnol. Bioeng.,* 2005.89(7): pp. 743-758.

Portner, R. and C. Giese, An Overview on Bioreactor Design, Prototyping and Process Control for Reproducible Three-Dimensional Tissue Culture, Chapter-2, Drug Testing in vitro: *Breakthroughs and Trends in Cell Culture Technology,* 2006.

Portner, R., S. N. Heyer, C. Geopfert, P. Adamietz and N. M. Meenen, Bioreactor design for tissue engineering, *J. Biosci. Bioeng.,* 2005. 100(3): pp. 235-45.

Prewett, T. L., T. J. Goodwin and G. F. Spaulding, Three-dimensional modeling of T-24 human bladder carcinoma cell line: A new simulated microgravity culture vessel, *Methods in Cell Sci.,* 1993. 15(1): pp. pp. 29-36.

Schleicher, M., G. Sammler, M. Schmauder, O. Fritze, A. J. Huber, K. S. Layland, G. Ditze and U. A. Stock, Simplified pulse reactor for real-time long-term in vitro testing of biological heart valves, *Ann Biomed. Eng.,* 2010. 38(5): pp. 1919-1927.

Schmidt, D. and Hoerstrup, S.P., Tissue-engineered heart valves based on human cells. *Swiss Medical Weekly,* 2006. 136(39-40): p. 618-623.

Schmidt D., J. Achermann, B. Odermatt, C. Breymann, A. Mol, M. Genoni, G. Zund, S. P. Hoerstrup, Prenatally Fabricated Autologous Human Living Herat Valves Based on Amniotic Fluid-Derived Progenitor Cells as Single Cell Source, *Circulation,* 2007.

Sutherland, F.W.H., Perry, T.E., Yu, Y., Sherwood, M.C., Rabkin, E., Masuda, Y., Garcia, G.A., McLellan, D.L., Engelmayr, G.C., Sacks, M.S., Schoen, F.J., and Mayer, J.E., From Stem Cells to Viable Autologous Semilunar Heart Valve. *Circulation,* 2005. 111(21): p. 2783-2791.

Syedain, Z. H. and R. T. Tranquillo, Controlled cyclic stretch bioreactor for tissue-engineered heart valves, *Biomaterials,* 2009. 30(25): pp. 4078-4084.

Vismara, R., M. Soncini, G. Talo, L. Dainese, A. Guarino, A. Radaelli and G. B. Fiore, A bioreactor with compliance monitoring for heart valve grafts, *Ann. Biomed. Eng.,* 2010. 38(1): pp. 100-108.

Zeltinger, J., L. K. Landeen, H. G. Alexander, I. D. Kidd and B. Sibanda, Development and characterization of tissue-engineered aortic valves, *Tissue Eng.,* 2001. 7(1): pp. 9-22.

Chapter VII

Concept and Development of Tissue Engineering Aortic Heart Valve

7.1. Introduction

Tissue engineering of heart valve is a novel technique set to develop a viable functional heart valve substitute from autologous cells (Sodian et al., 2000a; Bilodeau and Mantovani, 2006). The basic idea is to first transplant autologous cells onto a valve scaffold made of natural and/or synthetic biocompatible/biodegradable materials and then the seeded scaffold is conditioned in either *in vitro* or *in vivo* or both. During the conditioning, the polymer scaffold gradually degrades as the cells form their extracellular matrix (ECM) growths, given a functional autologous tissue-engineered heart valve for the same patient the cells were taken from (Sodian et al., 2000a). As outlined in Chapter 3, a functional tissue-engineered valve would possess substantial hypothetical advantages over the present heart valve substitutes, such as mechanical, glutaraldehyde-fixed xenografts, and homografts. These artificial valves are made from foreign body materials and as such are nonviable, and there are always the risks of thromboembolic complications, degeneration, calcification of the leaflets and infections. On the other hand, a tissue-engineered heart valve would in principle develop a tissue valve of the same physical characteristics that can be totally integrate and grow with the surrounding host. Such a concept will no doubt provide heart valve substitute capable of remodelling, which is particularly important paediatric patients (Wu et al., 2006; Morsi and Birchall, 2005; Morsi et al., 2004). As illustrated in figure 7.1 there are four main phases that need to be conducted in a systematic way, to achieve a tissue-engineered heart valve. These are:

Throughout these processes, the principle patho-physiological procedures are cell differentiation and coverage, ECM generation and structure, scaffold degradation and tissue remodelling (Novakovic and Freshney, 2006; Mendelson and Schoen, 2006).

As stated above, the concept of TEHV is based on the fact that autologous cells need to be harvested and isolated from *in vivo* and then seeded into the correct shape of the heart valve scaffold and then be conditioned *in vitro* or *in vivo*. Biomaterials are the second most important part of tissue engineering, and since they provide structural support to the cells, they need to be robust and flexible. Hence, design and optimization of the scaffold is very important for the success or otherwise of tissue-engineered heart valve. The scaffold should possess good mechanical and hemodynamic characteristics to stand the dynamic forces occurring on it when implemented *in vivo*. Moreover, the surface morphology and porosity to encourage the growth of tissue and the flow of the nutrients to it, degradation rate and appropriate surface chemistry to allow the tissue to be developed over it are also equally important (Morsi and Birchall, 2005; Atala, 2011; Maher et al., 2009; Yang, 2011; Brody and Pandit, 2007).

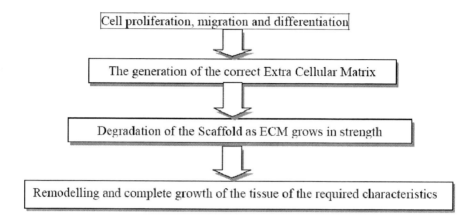

Figure 7.1. Tissue-engineered heart valve.

In tissue engineering of heart valve, scaffolds can be constructed from natural or synthetic materials or from a combination of both, as mentioned below. Decellularized, natural or biodegradable scaffold materials are used with heart valve cells such as fibroblasts, myofibroblasts and endothelial cells, to generate the ECM required. In literature, there are many types of decellular biomatrices as well as biocompatible polymers and/or biological degradable materials that have been identified and enveloped with various degrees of success, as scaffolds to support cell growth during tissue regeneration. In this chapter, the most significant progress and steps that have been reported in literature on the development of tissue engineering heart aortic valve using decellularization, biological/natural and/or polymer approaches are reviewed and discussed (Brody and Pandit, 2007).

7.2. Decelluarization Approach

As shown in Figure 7.2, aortic heart valve has a very complex configuration and structure. Owing to the heavy mechanical load on the aortic heart valves, decellularized valves are the preferred source of biomaterials; removal of cells from these valves results in identical shape of the valve and in reduced immunogenicity and possible complete intact of ECM, making the surface more compatible for the cells to grow onto.

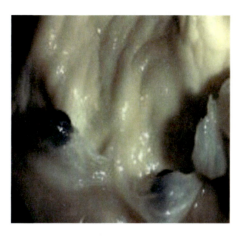

Figure 7.2. Photo of Aortic Sheep.

Table 7.1. Summary of decellularization methods and chaotropic agents

Methodology	Description	Comments of Extracellular matrix (ECM)
Physical stimulus		
Snap freezing	The creation of intracellular ice crystals, instigates the depletion of cells	This technique can disrupt and adversely affect the integrity of ECM
Action of mechanical forces	The pressure forces remove tissue and cells	There is definitely a side effect using this method as the use of high pressure can cause serious damage to ECM
Action of mechanical agitation	In general, agitation can instigate cell lysis and assist chemical exposure and depletion of cells	Aggressive agitation can trigger disorder of ECM as the cellular material is depleted
Chemical		
Alkaline; acid	This method solubilizes cytoplasmic mechanism of the cells and disrupts nucleic acids	The removal is not complete as this method eliminates removes GAGs from ECM
Non-ionic detergents		
Triton X-100	Although this method leaves protein to protein interactions unharmed, it does disrupt lipid–lipid and lipid–protein interactions	Inconsistency of findings but known to remove GAGs from ECM
Ionic detergents		
Sodium dodecyl sulfate (SDS)	It does solubilize cytoplasmic as well as nuclear cellular membranes; it can denature proteins	Depletes nuclear remnants and cytoplasmic proteins but it disrupts native tissue structure, damages collagen and eliminates GAGs
Sodium deoxycholate		It is known to cause more disruption to ECM than SDS
Triton X-200		Results in good cell removal when combined with zwitterionic detergents
Zwitterionic detergents		
CHAPS	Demonstrates characteristics of non-ionic and ionic detergents	Good cells depletion with some disruption to ECM

Table 7.1. (Continued)

Methodology	Description	Comments of Extracellular matrix (ECM)
Sulfobetaine-10 and -16 (SB-10, SB-16)		Good cell depletion and mild ECM disruption with Triton X-200
Tri(n-butyl)phosphate	This is a solvent organic, which is known to disrupt protein–protein interactions	Small effect on the mechanical properties and effectively removes cells, but it can result in loss of collagen content
Hypotonic and hypertonic solutions	This introduces cell lysis by osmotic shock	It is known to efficiently deplete the cellular components
Methodology	Description	Comments of Extracellular matrix (ECM)
Chelating agents EDTA, EGTA	An agent that binds divalent metallic ions, thus disrupting cell adhesion to ECM	No isolated exposure, and it is typically used with enzymatic methods (e.g., trypsin)
Enzymatic		
Trypsin	To adhere peptide bonds on the C-side of Arg and Lys	Lengthy duration can disrupt ECM structure resulting in removal of laminin, fibronectin, elastin, as well as GAGs
Endonucleases	Causes catalyzation of the hydrolysis of the interior bonds of ribonucleotide as well as deoxyribonucleotide chains	Hard to remove from the tissue and could cause an immune response
Exonucleases	Catalyzation of the hydrolysis of the terminal bonds of ribonucleotide and deoxyribonucleotide chains	Good removal of cells

With this approach, donor- or animal-derived valves (allogenic or xenogenic) are drained of cellular antigens and are subsequently used as a scaffold material to regenerate new tissues. Adapting this approach will ensure that the valves are free of endothelial and interstitial cells, as well as cell debris, leaving the morphology of the ECM well preserved. Generally, the depletion of valve cells involves the use of Trypsin/ Ethylenediaminetetraacetic acid (EDTA), followed by sodium dodecyl sulphate (SDS) washing, though other techniques are also used. In literature, various decellularization techniques have been proposed that will remove cells and do minimal or no harm to the ECM. These methods are typically classified as physical, chemical and enzymatic methods and are briefly discussed in Table 7.1 (Gilbert et al., 2006).

Researchers in general considered a combination of the above-listed techniques to improve the decellularization process. For example, cell membrane would be lysed initially using any of the physical methods (like sonication, mechanical pressure, freezing or thawing) or by the use of ionic solutions, then the cellular components would be separated from the ECM by enzymatic treatments. However, in general, the effectiveness of decellularization technique is dependent upon the tissue in question, as one decellularization method could be successful for one type of tissue but might not be as effective other tissues. In addition, the source of the tissue has an effect on the efficiency of decellularization procedure as well, as observed in the work of Schenke-Layland et al., (2004). Still, care should be taken with any decellularization method to remove all the residual chemicals prior to cell seeding and conditioning, as these chemicals could be toxic to the cells and adversely affect cell regeneration (Gilbert et al., 2006). In literature, there are numerous investigations using decellularization approach and utilizing different decellularization techniques to regenerate a tissue engineering heart valve, however, only a selected numbers of these techniques are discussed here.

In the early study of Curtil et al. (1997), *freeze-drying technique* has been used to produce porous scaffold, which was subsequently seeded with low-density human dermal fibroblast and umbilical endothelial cells. Although both cells formed confluent layers on the scaffold after conditioning, the generated ECM collapsed, and the valve failed due to build up ice in the valve leaflets. Later, Bader (Bader et al., 1998) used decellularized porcine aortic valves (pig) and ovine (sheep) pulmonary valves for *in vitro* and *in vivo* tissue regeneration. In both of these experiments, native cells were successfully removed from the biomatrices, and the ECM organization of the biomatrices was preserved and subsequently repopulated with cultured cells. However, the findings of the *in vivo* trials of similar study by Steinhoff et al., (2000) showed thickening and calcification of the valve leaflets but without any apparent loss of function or pulmonary regurgitation, which was considered to be very positive at that time.

The decellularization approach was also adapted by O'Brien et al. (1999) who chose to not seed biomatrices with cells before grafting the valve in female sheep for six months. The results showed no regurgitation; calcification or gross abnormalities and some cells coverage similar to the cryopreserved, cellularized, allogenic sheep aortic valve were observed. Based on these positive results at the time, the commercial SynerGraft pulmonary replacement valve was proposed for implementation in humans, and in October 2000, CryoLife Inc., obtained a CE mark to supply this type of technology in Europe. However, when this valve was model-grafted in a three-year-old male child in Norway, unfortunately there were disastrous consequences, as the child recipient died after a short period of time (Brody and Pandit, 2007).

Steinhoff et al., 2000 decellularized lamb pulmonary valves by immersing the valves into a solution of 0.05% trypsin, and 0.02% EDTA in controlled ambient parameters under constant mechanical shaking. It was observed that there was extracellular matrix ECM of valve after decelluraization. Subsequently, the valves were seeded with a combination of endothelial and myofibroblast cells and both seeded and unseeded valves were implemented into lambs. Histology analysis of the decellularization procedure revealed a patchy incomplete seeding of endothelial and myofibroblast cells on the surface of the valves, but after 12 weeks, implantation of valves confluent layer of endothelial cells was formed on all unseeded group. Moreover, microscopic valve morphology observation of all the valves showed different degrees of subvalvar calcification of remnant muscular tissue on all valves examined. Additionally, vulvar calcification was only presented on one valve belonging to unseeded group, and no calcification was noticed for the supravalvar conduits of unseeded and TE valves.

In another study by Samouillan et al. (1999), porcine aortic valves were decelluraized by two detergent techniques, and the preserved extracellular matrix (ECM) of each was analysed by characterization of elastin and collagen. The valve sample was treated with the aid of 1% (w/w) sodium dodecyl sulphate (SDS), whereas the other sample was treated with 1% (w/w) triron x-100 and cholate (TRI-COL). Differential scanning calorimetriy (DSC), and Thermally Stimulated Current (TSC) were utilized to analyse the structural and dynamic parameters of the major protein within the ECM. It was reported that when the results are compared with the non-treated valve, no difference was noted at the level of localised motion of collagen and elastin between SDS and TRI-COL preserved valves. However, although both samples showed no disruption of ECM structure, SDS and TSC studies at high temperature showed significant differences of thermal transitions and dielectric relaxations parameters between the two treated valves, with clear destabilising effect of SDS detergent. This, in return, implies that TRI-COL is better method to be used in the preservation of ECM.

Zeltinger et al., (2001) seeded decellularized aortic porcine heart valve with human neonatal dermal fibroblasts (DmFbs). The authors used Tris-Buffer, phenylmethylsulfonyl fluride (PMSF), Protease inhibitor to break the native cells from the valve and extracted some cytoplasmic elements, followed by several treatments with DNase I, RNase A, Phospholipases A2, C, and D, Trypsin protease to digest nucleic acid and cellular membranes. Histology analysis of the decellularized valve showed successful depletion of endocardial cells and myofibroblasts, and ECM architectures such as fibrosa, spongiosa and ventricularis were well preserved. Two types of samples (whole valve, and leaflets) were seeded and cultured in pneumatic bioreactor (PB) for one, two, four, six and eight weeks and subsequently analysed. The results revealed that cells attached to the surface of decellularized porcine matrix from day one through week eight with recellularization density reached about 45% of native porcine leaflets after eight weeks. However, disruption of cells on the valve surface in the regions of the leaflets corrugation after 24 hours in bioreactor was observed, and cells didn't attach to the surface of the leaflet commissures.

Booth et al., (2002) carried out a study to improve the process of producing a porcine scaffold with decellular matrix, which was used in the development of a tissue-engineered heart valve containing autologous cells. The decellularization process was performed by using different type of detergents known as Triton X-100, sodium dodecyl sulphate (SDS), sodium deoxycholate, MEGA 10, TnBP, CHAPS and Tween 20. The total decellularization protocol resulted in 72 hours and followed by the reservation of the whole cells and cell fragments. The authors stated that using these techniques of decellularization, the main structural components of the valve matrix have been preserved and also seem to be undamaged from the effects of chemicals.

In 2002, Kim et al., investigated three different decellularization techniques in order to find the most effective method in terms of producing decellularized valve xenograft. Freeze-thawing, Triton and NaCl-SDS treatments were used for decellularization of porcine pulmonary leaflets. After that, the decellularized xenograft valve leaflets were seeded with endothelial cells, which were obtained from the jugular vein of a goat. It was reported that complete removal of cells in the leaflets was successfully achieved by NaCl-SDS treatment compared to the Triton and freeze-thawing methods.

Later, Dohmen et al., (2003) seeded ovine autologous vascular endothelial cells (AVECs) onto decelluarized porcine pulmonary. Deoxycholic acid-treated valves seeded in static culture for seven days and then were implanted in juvenile sheep. Subsequent analysis of the harvested valves collected after seven days, three months and six months showed normal ingrowth on both sides of anastomoses, without encapsulation of the leaflets. Moreover, good mechanical properties of the implanted valves were observed with no sign of tearing, perforation, cusp deformation, retraction of the cusp or hardness with good coaptation after three and six months, respectively. In addition, light microscopy results showed that there was no fibrous tissue overgrowth at any stages for all collected valves. Moreover, a confluent monolayer of AVECs formed at valves wall and leaflets was observed. Still, fibroblasts started to develop after three months of implantation with histological evidence for collagen production.

Leyh et al., (2003) aimed at determining whether scaffolds made from decellularization of xenograft are immunogenically active, can potentially transmit the disease from donor to host and whether porcine endogenous retrovirus (ERVs) is capable of infecting human cell lines. To achieve this, the porcine aortic valve was decellularized by trypsin/EDTA detergent, seeded with ovine myofibroblasts, and subsequently coated with endothelial cells. Results from the harvested valves after six-month implantation in sheep showed that even after decellularization of the valve, 2% of native DNA was still detectable. However, after six months of implantation, no sign of porcine (ERVs) was observed, indicating that chemical treatment with trypsin/EDTA was sufficient to prevent transmission of porcine (ERVs). However, further *in vivo* trials are called for to validate this finding.

Spina et al., (2003) reasoned that development of durable bioprosthesis could be achieved by the use of extracellular matrix (ECM) scaffold separated from valvulated conduits. In their study, Triton X-100 and cholate (TRI-COL) or N-cetylpyridinium (CPC), respectively, treated porcine aortic valves. Removal of majority of cells and retaining of endothelium membranes, ECM texture and mechanical properties were achieved by the treatments used. Moreover, the extracellular matrix (ECM) of scaffold treated by TRI-COL appeared to be very similar to natural heart valve with very low possibility of calcification.

In their study, Bertipaglia et al., (2003) treated porcine aortic valves using ionic and non-ionic detergents such as Triton X-100, NaDC and benzonase. Then the decellularized valve leaflets were seeded by aortic valve interstitial cells (VIC), and after 15 days of seeding period, grafted cells were observed to be incorporated with the bioscaffold. The authors achieved excellent histochemical and ultrastructural preservation.

The aim of the study accomplished by Stamm et al., (2004) was to produce matrix/polymer hybrid valves in order to prevent thrombogenic reactions. The porcine aortic valves were first treated with Trypsin and then deep-coated in a solution of poly (4-hydroxybutyrate) (P4HB), Poly (3-hydroxybutrate) (P3HB), and poly (3-hydoxybutrate-co-4-hydoxybutrate), respectively. Subsequently, the decellularized heart valves were implanted into lamb model in pulmonary and aortic positions, and it was observed that the residual antigenicity associated with xenograft valve can be sorted out by pre-treatment (coating) with biodegradable polymer. Furthermore, evaluations of constructed valve showed excellent biologic and biomechanic characteristics of matrix/polymer hybrid valve, which can be described as promising development for valve replacement application. However, all the valves implanted in pulmonary position failed.

Schenke-Layland et al., (2004) proposed to analyse the differences between original and tissue-engineered valve leaflets in terms of extracellular matrix (ECM), cell location and cellular phenotypes. Porcine pulmonary valves were decellularized by Trypsin/EDTA treatment and then seeded by ovine endothelial and myofibroblast cells and condition *in vitro* for up to nine days. The authors stated structural similarity between native and tissue-engineered valve leaflets were present, although the valvular cell phenotypes of original and tissue-engineered leaflets pointed out a considerable difference.

Later, Rieder et al., (2004) investigated the effect of decellularization methods on valve leaflets and root matrices. The authors used a combination of Triton X-100 and sodium-deoxycholate and washing process (with M-199 medium and an enzymatic digestion with DNAse/RNAse) in the decellularization of xenogenous constructs. The researchers achieved an excellent scaffold, free of any histologically detectable xenogenous cells and observed significant variations with respect to the productivity of total xenogenous cell removal and the receptiveness to human cell attachment to the decellularized valve. Still, Rieder's group also used sodium dodecyl sulphate (SDS) and found that although a biological matrix free of any visible xenogenous cells was produced, it was impossible to have sufficient coverage of cells on the porcine aortic or pulmonary root specimens with human endothelial cells or myofibroblasts, as huge cell lysis developed within 24 hours of cultivation. This was attributed to "probably" higher levels of residual SDS remaining in the vascular tissue. They further stated adequate cultivation under shear stress would provide the correct environment for a good adherence of seeded cells to acellular matrix surface, and cell cultured under continuous flow environments would not have substantial influence on the biocompatibility of xenogenous matrices.

Grauss et al., (2005) utilized two different techniques of treatment to regenerate decellular matrix scaffold and examine the variations in extracellular matrix constitution and components. Porcine aortic valves were decellularized by using detergent treatment (Triton X-100) and enzymatic technique (Trypsine), and it was reported that use of chemical methods for decellularization process resulted in changes in extracellular matrix composition. Moreover, in both treatments, the content of collagen was lower, and GAGs content was nearly zero.

Kasimir et al., (2005) used 0.5% Triton X-100, 0.05% SDS, 0.05% Igepal CA-360 and RNase to decellularize porcine heart valve conduits and subsequently seeded with endothelial cells. It was reported that the process achieved a complete removal of cells, and the collagen and elastin contents were maintained. Moreover, no platelet activation occurred, and endothelial cells were able to grow and join.

Later, Lichtenberg et al., (2006) decellularized ovine pulmonary valve, using 0.5% NaDC and 0.5%sodium dodecyl sulphate (SDS) and subsequently seeded it with endothelial cells (ECs). The three seeded pulmonary valve models were tested in bioreactor under the correct physiological conditions for 36 hours and were subsequently examined from morphology and viability of cells point of view. It was observed that the generated monolayer of EC appeared to cover the luminal surface of the conduit of the valve as well as both sides of the valve leaflets. Moreover, the issue matrix developed showed comparable values of collagen, elastic fibre, and glycosaminoglycan (GAG) to native valve. Still, the SEM observations indicated that ECM had a 3D structure network and completely maintained the basement membrane in the inner surface of pulmonary wall, as well as both sides of the valve leaflets with sufficient cells attachment and stability. It was also emphasized that to ensure a good regeneration of stable cell matrix and cell connectivity, the correct cell culture parameters such as temperature, nutrition, pO_2, pCO_2 and pH must be maintained. Moreover, controlling and gradually introducing the flow dynamics is equally important in maintaining EC sustainability and viability.

Moreover, Tudorache et al., (2007) examined the effect of three different decellularization protocols on the structure integrity and surface morphology of porcine pulmonary valve. Cells were removed from porcine valve by SD, SDS, and trypsin/EDTA decellularization techniques. The findings indicated that both detergent-treated valves appeared to preserve ECM, with all cells completely detached, and enzymatic-treated valves, on the other hand, had ECM disruption.

It was then argued that the central awkwardness of all decellularization approaches is immunogenicity, thrombogenicity, and ECM disruptions. In this respect, Liao et al., (2008) carried out studies to analyse the impact of different treatments. The authors treated porcine aortic valves with sodium dodecylsulafte (SDS), trypsin, and triton X-100, and observed ECM disruption and early degradation. It was reported that SDS-treated valves maintained the critical mechanical properties and resembled the ECM of native valve. On the other hand, treated porcine pulmonary valves with A) 1% sodium deoxycholate (SD), B) 1% sodium dodecyl sulfate (SDS), C) 0.05% Trypsin /0.02% EDTA, D) 0.1% Trypsin/0.02% EDTA (Zhou et al., 2010). Histological analysis showed cells removed successfully with all decellularization protocols, while ECM disruptions were noticed for all groups except valves treated with SD. Furthermore, all four groups showed increase human immune response and thrombogenicity (Zhou et al., 2010).

In the study of Seebacher et al., (2008), decellularization process of porcine pulmonary valve conduits was performed by using enzymatic treatment (RNase and DNase). The removal of complete cells was achieved. It is also stated that the difference between decellularized porcine conduit tissue and human homograft tissue was significant.

Recently, an interesting investigation was carried out by Hong et al., (2009), where the decellularized porcine aortic heart valve scaffold was coated with basic fibroblast growth factor (bFGF)/chitosan/poly-4-hydroxybutyrate (P4HB) using electrospinning technique and subsequently seeded with mesenchymal stem cells (MSCs). A confluent layer of MSCs was observed on the scaffold, and the hybrid valve leaflets (bFGF) showed good cells attachment, growth, and adequate mechanical strength. De Cock et al., (2010) conducted similar study using a decellularized porcine aortic heart valve structure combined with heparin, and heparin/bFGF (deposited layer by layer) technique was investigated. The results showed that bFGF scaffold in general stimulates cells proliferation, which again is line with the results of Hong et al. (2009). Zhou et al. (2010) adapted the same treatment of Tudorache et al. (2007), namely SD, SDS, and trypsin/EDTA techniques on porcine aortic heart valves. The authors found, however, that only sodium deoxycholate (SD)-treated valve preserved ECM with maximum cells removal, whereas the rest did not.

In the study of Lu et al., (2010), decellularization process of porcine aortic valves started with Trypsin treatment and followed with EDTA, Triton X-100, SD, Rnase and Dnase, which can be formulated as a composite of chemical and enzymatic techniques. Endothelial cell (ECs) and valve interstitial cells (VICs) were implemented onto decellularized porcine aortic valve scaffold. The results indicated that NDGA cross-linked decellularized scaffolds had several advantages over non-cross-linked valve scaffolds in terms of tensile strength, enzymatic hydrolic resistance and store stability. In addition, NDGA cross-linked porcine aortic valves provide excellent mechanical properties and cytocompability. The findings indicated that porous scaffold was achieved without any disruption of ECM structure, and the endothelial cells (ECs) generated showed a good ability of attaching to NDGA cross-linked scaffold and retaining the morphology of the surface.

More recently, Deng et al., (2011) used hybrid valve (decellularized porcine aortic heart valve coated with PEG nanoparticles) to investigate the effect of decellularization methods. The authors used porcine aortic leaflets, which were first treated by trypsin/EDTA, followed by triton X-100, RNase, and DNase treatments, which were named as "simple scaffolds." Still, the leaflets that were combined with PEG, nanoparticle via a coupling reagent, carbodiimide, were called "modified scaffolds," and the ones that were loaded with TGF-β1 were named "delivery scaffolds." In general, however, it was observed that ECM preserved better on delivery scaffolds, with clear evidence of confluence layers of cells formed on the surface of scaffolds. Moreover, the maximum stress load on delivery and modified scaffolds were almost 70% more than that of simple scaffolds.

Table 7.2 gives a summary of decellularization approaches to construct tissue engineering heart valve as discussed by Brody and Pandit (2007) and many other researchers.

7.3. Scaffolds Made from Biocompatible Materials

Initially, various researchers were encouraged by the physical characteristics of biological materials, such as excellent seeding properties and controlling of degradation, and used different types of biological biodegradable scaffolds to regenerate tissue of heart valves particularly leaflets. However, due to number of limitations related to mechanical strength of the biological scaffolds, other biocompatible synthetic materials were explored alone and with supplement of biological materials as possible constructs. With this approach, the scaffold is constructed from either biological and/or biocompatible biodegradable synthetic materials. Synthetic materials have a number of advantages including ease of manufacturing and good mechanical integrity. However, design of the surface morphology of scaffold made from synthetic materials can be complex, and controlling degradability is equally difficult. In the following section, a brief discussion of a selected number of investigations using biological and synthetic materials or combination of both to construct scaffolds for tissue-engineering applications is presented and discussed.

7.3.1. Biological Materials

As observed, decellularized valves are not ideal biomaterials. Hence, other natural materials like collagen, fibrin, chitosan, hyaluronic acid and so on were studied as possible scaffolds. Collagen biomaterials are very commonly used, as they cause hardly any antigenic response as compared to other naturally available biomaterials, and they have also been approved by United States Food and Drug Administration (FDA) (Yarlagadda et al., 2005).

Moreover, when these biomaterials are implanted *in vivo,* they gradually degrade under the influence of lysosomal enzymes, and their degradation rates depends upon the intra-molecular cross-linking amongst the molecules, which can be readily controlled. In this section, a review is given on the use of biological materials for tissue engineering of heart valve.

Vesely et al. carried out a study in 1998, aiming at establishing the link between elastin and collagen structures within the ECM of porcine aortic valve leaflets (Vesely et al., 1998). The authors found that minimum forces were associated with the elastin structures, with an elastic modulus of only 0.05% of the whole tissue. They isolated elastin from porcine aortic valve leaflets and mechanically tested elastin tissues incorporated with them and compared the results with the original leaflet layers. It was found that elastin influence on tissue mechanics was noteworthy only at low strains and varied in magnitude between the fibrosa and the ventricularis.

Table 7.2. Summary of decellularization approaches to construct tissue engineering heart valve

Reference	Scaffold Material	Decellularization Treatment	Cell source and cell seeding observation	General observation and generation of ECM
Curtil et al., 1997	Porcine Pulmonary valve,	Physical Method/ Freeze-drying	Human dermal fibroblast and umbilical vein endothelial cells Condition *in vitro* for almost three weeks	Low density of fibroblast cells Enough cells attachment Cells metabolically active Endothelial cells formed a confluent layer
Bader et al., 1998	Porcine aortic heart valve	Using a non-tanning detergent extraction procedure / Treated with 1% Triton X-100 and 0.02% EDTA, RNase and DNase, and washing with PBS	Human Endothelial cells Condition *in vitro* for three days	Scaffold was micro-porous only partially EMC produced consisted of loosen original collagen and elastin fibre Observed confluent layer of cells on the surface
O'Brien et al., 1999	Female sheep valve	After six months, following the implantation	Selected to not seed biomatrices before *in vivo* female sheep for six months trials,	No sign of abnormalities, no pulmonary regurgitation or even calcification or Host sheep cells re-populated the scaffold with no significant difference from cryopreserved.
Steinhoff et al., 2000	Ovine Pulmonary Valves	Decellularization Process/ Treated with 0.05% Trypsin and 0.02% EDTA, washed with PBS	Ovine myofibroblast cells seeded for six days, followed by seeding endothelial cells for two days, followed by *in vivo* implementation for 12 weeks.	Excess extra cellular matrix formation led to leaflets becoming thicker *In vivo* study, confluent layers of endothelial cells on the scaffold achieved after four weeks, *In vitro* study, cultured cells didn't generate completely and was patchy.
Samouillan et al., 1999	Porcine aortic heart valve	Chemical Method-Ionic detergent Treatment/ Treated with 1% (w/w) sodium dodecyl sulphate (SDS) and Tri-COL	Not seeded N/A	Both treatments didn't disrupt ECM, Treated valves showed similar level of localised motions of elastin and collagen for both treatments
Reference	Scaffold Material	Decellularization Treatment	Cell source and cell seeding observation	General observation and generation of ECM
Zeltinger et al., 2001	Porcine aortic heart valves	Enzymatic Treatment/ Treated with Tris-Buffer, Phenylmethylsulfonyl fluoride (PMSF), Protease inhibitor, followed by DNase I, RNase A, Phospholipases A2, C, and D, Trypsin protease.	Human neonatal dermal fibroblast cells (DmFbs) and *in vitro* for eight weeks	ECM, i.e., Fibrosa, Spongiosa, and ventricularis was almost preserved. After the conditioning of eight weeks, human cell density reached to 45% of the native porcine cell density. Collagen expression reported
Booth et al., 2002	Porcine Aortic Heart Valve	Detergent treatments/Treated with Triton X-100, SDS, sodium deoxycholate, MEGA 10, TnBP, CHAPS and Tween 20	Autologous Cells 72 hours	Sustained significant amount of structural proteins of matrix

Table 7.2. (Continued)

Reference	Scaffold Material	Decellularization Treatment	Cell source and cell seeding observation	General observation and generation of ECM
Kim et al., 2002	Porcine Pulmonary valves	Freeze-thawing, Triton and NaCl-SDS treatment	Autologous vascular endothelial cells (AVEC)	The valves treated with NaCl-SDS showed better results in terms of efficacy of decellularization Well-preserved collagen structure
Dohmen et al., 2003	Porcine Pulmonary valves	Treatment/ Treated with 0.1% Deoxycholic acid	Ovine Autologous vascular endothelial cells (AVECs) *In vivo*, seeded in static culture for seven days, then implanted up to six months	Good mechanical properties observed, No perforation, cusps tearing or deformation observed after six months. Continents in the middle of the cusps showed fibroblast cells growth between three to six months. Confluent layer of cells observed at commissures
Leyh et al., 2003	Porcine pulmonary heart valves	Decellularization Process/ Treated with Trypsin/EDTA	Seeded with ovine myofibroblasts and coated with endothelial cells Condition *in vitro* for 18 days, then *in vivo*, up to six months.	Recellularized valve had similar collagen, elastin and GAG content to native valve
Spina et al., 2003	Porcine aortic heart valve	Freeze-Drying and treated by TRI-COL, and 0.1% N-cetylpyridinium chloride(CPC)	Human dermal fibroblast cells *In vivo*	ECM obtained from TRI-COL-treated valve had similar structure to original valve and had a very low calcification. Good cell attachment and grow for TRI-COL treated leaflet
Bertipaglia et al., 2003	Porcine aortic valve	Ionic, non-ionic detergents Treated with 1% Triton X-100, NaDC, benzonase	Porcine aortic valve VICs *In vitro* conditioning for two weeks	Good histochemical and ultrastructural preservation Recellularized valve had Collagen and elastin
Stamm et al., 2004	Porcine aortic valves	Enzymatic Treatment/ Treated with 0.05% Trypsin and coated with co-polymer of 18% P4HB and 82% of P3HB	Rat fibroblast *In vivo*, up to three months	Not reported Collagen deposition and endothelial cell lining observed
Schenke-Layland et al., 2004	Porcine Pulmonary valve	Enzymatic Treatment/Treated with Trypsin, EDTA	Ovine endothelial cells and myofibroblast cells Static *In vitro*, up to nine days	Similar structure to native porcine, and ovine valves achieved, Confluent layers of endothelial cells on the scaffold formed
Rieder et al., 2004	Porcine aortic and pulmonary valve	Chemical & Enzymatic Treatments withTriton X-100, NaDC, RNase, and DNase	Human endothelial cells and myofibroblast *In vitro*, endothelial cells up to five days and Myofibroblast cells, up to ten days	Both endothelial and myofibroblast cells formed confluent layer on the scaffold

Reference	Scaffold Material	Decellularization Treatment	Cell source and cell seeding observation	General observation and generation of ECM
Grauss et al., 2005	Porcine aortic valve	Chemical & Enzymatic Treatments/ Treated with Triton X-100,EDTA,Gentamacin,DNase and RNase Enzymatic Treatments/ Trypsin, EDTA, Gentamicin, DNase and RNase	Porcine aortic valve VICs Static condition in *In vitro*, up to four weeks	Collagen content reduced and GAGs almost completely lost in both treatments, Endothelial cells reached to 80% confluency of Triton X-100 treated valve.
Kashmir et al., 2005	Porcine heart valve	Chemical & Enzymatic Treatments/ Treated with 0.5% Triton X-100, 0.05% SDS, 0.05% Ideal CA-360 and RNase	Human endothelial cells	Complete cell removal, no changes on collagen and elastin content. Endothelial cells grew and became confluent, no platelet activation reported
Lichtenberg et al., 2006	Ovine Pulmonary valve	Chemical Treatment (use of ionic detergents)/ Treated with 0.5% NaDC and 0,5% SDS	Ovine endothelial Cells	Sustained the ECM fibbers with the basement membrane conserved, Decellularized PV showed lower tissue resistance to tension than native valve, in both longitudinal and circumferential direction. Endothelial cell monolayer on a scaffold showed sufficient adhesion in bioreactor and became confluent
Tudor ache et al., 2007	Porcine Pulmonary valve	Chemical & Enzymatic Treatments/Decellularized with 1% sodium deoxycholate (SD), 1% Sodium deadeye sulphate (SDS), 0.05% trypsin/0.02% EDTA	Not Seeded	Both detergent methods completely removed all cells with protection of ECM integrity. Enzymatic treatment showed disrupted ECM
Liam et al., 2008	Porcine aortic heart valve	Chemical & Enzymatic Treatments/ 12 valves were treated with three methods, 1) by SDS with RNase and DNase 2) by Trypsin/EDTA with RNase, 3)Triton X-100 with DNase and RNase	Not Reported	The three decellularized valves displayed an increase in leaflet extensibility and Triton X-100 had the maximum on the radial extensibility. ECM had little damage and disruption. Spongiosa layer depleted by Triton X-100 and Trypsin treatment. ECM of SDS resembled the native valve For Triton X-100 and try sin might have early degradation
Debaucher et al., 2008	Porcine Pulmonary heart valve conduits	Enzymatic Treatment/ treated with rib nuclease (100μg/mL) (RNase) and deoxyribonucleic (150 IU/mol) (DNase)	Not reported	All cells removed completely, Similar tension and failure strain both in longitudinal and transverse direction to Native valve, but higher than human homograft.

Table 7.2. (Continued)

Reference	Scaffold Material	Decellularization Treatment	Cell source and cell seeding observation	General observation and generation of ECM
Hong et al., 2009	Porcine aortic heart valve coated with Chitosan/P4HB and bFGf/Chitosan/P4HB	Enzymatic Treatment/ Leaflets treated with try sin/EDTA	Rat Mesenchymal stem cells	Hybrid valve leaflet (bFGF) has a significant improvement of strength. Good cell attachment and growth
Zhou et al., 2010	Porcine Heart Valve	Chemical Treatment (use of detergents) & Enzymatic Treatments/with A) 1% sodium deoxycholate (SD), B) 1% sodium dodecyl sulfate (SDS), C) 0.05% Trypsin /0.02% EDTA, D) 0.1% Trypsin/0.02% EDTA	Not Seeded	Only group A (treated with sodium deoxycholate) preserved ECM structure with optimum cell removals.
Lu et al., 2010	Porcine Aortic Heart valve	Chemical & Enzymatic Treatments/Decellularized with Trypsin solution(0.25%) followed by (0.0% Triton X-100 0.5% SD and 0.02% EDTA) and RNase and DNase	Umbilical vein endothelial cell (ECs) And Valve interstitial cells (VICs)	NDGA-Cross-linked valve resemble the natural valve. Porous scaffold without any disruption of ECM structure achieved. Demonstrated ECs can attach to NDGA-cross-linked scaffold with nearly 100% viability Only enzymatic degradation for 24hr observed
De Cock et al., 2010	Porcine Aortic Heart valve coated with bFGF and heparin	Freeze-Drying (Lyophilization) The first stage: mild alkaline hypotonic(PH8)and Hypertonic buffered solution (ph8), with phenylmethylsulfonyl fluoride (PMSF), streptomycin/penicillin solution, and butylated hydroxyanisole (BHA), The second stage, treated by DNase I, RNase A, Trypsin and phospholipases A2, C and D, followed by filling of bFGF and heparin	Human dermal foreskin fibroblast cells	Pore size of 1 to 15µm achieved, Collagen content was measured to be 85% Demonstrated that loaded bFGF on the scaffold stimulates proliferation, in contrast heparin did not influence cell proliferation

Reference	Scaffold Material	Decellularization Treatment	Cell source and cell seeding observation	General observation and generation of ECM
Deng et al., 2011	Porcine aortic heart valve coated with PEG nanoparticles	Chemical & Enzymatic Treatments/Leaflet immersed in (penicillin 100mg/mL, streptomycin 0.4 mg/mol, amphotericin B 0.5 mg/mol) and then treated by solution of 0.05% try sin and 0.02% EDTA, followed by solution of 1% Triton X-100 and 20mg/mL of DNase and 200μm DNase.	Rat myofibroblast cells	Hybrid scaffolds with Nano PEG coated achieved. Tensile strength, Maximum stress, strain increased on coated scaffold compared to simple scaffold Good cell attachment and growth observed, DNA analysis proved superior recellularization.

Moreover, the elastin tended to dominate the distensibility curves of the radial ventricularis and contributed very little in the fibrosa. Hence, the observations of this study suggested that the function of elastin in the aortic valve leaflet is to maintain the required collagen fibre structures on pressure and forces the fibres into these setup once outer forces have been removed. These findings of Vesely's group have assisted the understanding of the tissue structure of heart valve leaflets.

Later, Rothenburger et al. (2002) carried out a study aiming at determining the role of proteoglycans in porcine heart valves. Scaffold made from type I collagen was seeded with myofibroblasts and subsequently endothelial cells on the top of preconditioned myofibroblasts scaffold. The newly formed tissue underwent histological and immuno-histochemical analysis, and the results showed that the large-sized proteoglycan (154 nm) was placed in the outer region of the collagen bundles in a rarely structured extracellular matrix compound. The proteoglycan of smaller size (46 nm) was observed to be aligned along the collagen bundles at gaps of 60 nm, whereas the middle -sized proteoglycan (56 nm) was established on the cell surface of myofibroblasts. The glycosaminoglycans, on the other hand, contained 80% chondroitin and dermatan sulfate and only 20% heparan sulfate. It was, therefore, concluded that proteoglycans could perform an imperative function in the integrity of cardiovascular tissues. Furthermore, this study showed that a heart valve-like tissue with proteoglycans, similar in property to the regeneration and dissemination and proteoglycans of native heart valves, could be manufactured (Rothenburger et al., 2002).

Consquently, Taylor et al., (2002) carried out an *in vitro* experiment to determine the appropriateness of a 3D biodegradable collagen sponge for cell viability, proliferation and phenotype of cultured human cardiac valve interstitial cells (ICs). A 3D collagen sponge scaffold was seeded with interstitial cells, cultured from cardiac valve tissues and subsequently stained with immunofluorescence. A panel of antibodies was used to determine cell phenotype, whereas the cell viability was assessed using a dye-based cell proliferation assay and cell death by lactate dehydrogenase measurements. As expected, the authors reiterated the fact that collagen is a suitable biodegradable scaffold and quite efficient in sustaining viable valve ICs and appears to enhance the capacity of the cell to express its original phenotype. Moreover, it was also observed that the use of fibrinencourages cellular migration, proliferation and matrix production through the release of platelet-derived growth factor and transforming growth factor-β (Brody and Pandit, 2007).

Later, Ramamurthi and Vesley (2005) employed continuous sheets of elastin and placed it on the top of biodegradable hydrogels (hylans) that possess cross-linked hyaluronan, a glycosaminoglycan to fabricate the 3D construct. In this work, the authors cultured neonatal rat aortic smooth muscle on scaffold of hylan gels that possesses micro-textured surfaces, as well as on plastic, and periodically analysed the components of the extracellular matrix (collagen and elastin). They found that the hylan substrates induced rapid proliferations of the cells over extended periods (about four weeks) compared to those cultured on plastic (two to three weeks). However, the amount of elastin derived from hylan-based cell cultures was found to be always greater (by about 25%) than that obtained from cells cultured on plastic. However, it was also noted that there was no significant difference observed between the two substrates beyond the first two weeks of culture, when the elastin content was normalized to the cell DNA content. On the other hand, at periods of more than two weeks of culture, cells conditioned on hylan gels generated 56% less amount of collagens/nanograms of DNA than that from the cells cultured on plastic. Moreover, a unique type of matrix layer, rich in elastin, was deposited at the hylan-cell interface by the cells gown on hylan gels, and this elastin demonstrated structure of organised perforation sheets similar to that of ventricularis layer porcine aortic valve leaflets. This led the authors to conclude that hylan gels should be considered as ideal substrates for inducing elastin synthesis in culture for obtaining structures that are similar to the elastin matrix of the native aortic valve. Subsequently, these types of elastin-hyaluronan composites may find favourable use in tissue engineering of replacements for the glycosaminoglycan- and elastin-rich layers of the native aortic valve cusp (Ramamurthi and Vesely, 2005).

Shi et al., (2006) proposed a method to create the effective structures of the aortic valve leaflets individually *in vitro* and, once they were established, to incorporate them into the valve structures. The authors constructed collagen fibre bundles scaffold using the source of directed collagen gel contraction together with neonatal rat aortic smooth muscle cells as well as acid-soluble type I rat-tail tendon collagen. The collagen gels were moulded into rectangular with porous end containers that force the gels to move longitudinally but permit contraction to take place in crosswise direction. Sets of such configurations were located in direct interaction with each other and conditioned further to verify whether they integrated to form continuous tissue or not. The observations showed that highly compressed and aligned collagen fibre bundles were constructed after eight weeks of culture. However, it was noticed that the shape of the mould could significantly affect their mechanical property, as the results showed that multi-branched fibre composes were the weakest, but the elongated constructs were the toughest. Moreover, the authors observed that integration of collagen constructs occurred *in vitro* and that the fabrication of a composite structure *in vitro* could be achieved. In the meantime, Neidert et al., have carried out an another study to construct a scaffold that is capable of mimicking the native heart valve using collagen type 1 and neonatal human dermal fibroblasts (Neidert and Tranquillo, 2006). The authors observed similar collagen fibre alignment and geometry with native cusp and root. However, the authors pointed out that although gross collagen fibre alignment was observed to be similar to native valves, the failure mechanism was different, suggesting different microstructures of the material used.

To address some of the above-mentioned limitations of collagen, Jockenhoevel's group (Jockenhoevel et al., 2001b) used an injection-moulding technique to construct a complex 3D gel tissue structures and changed the thickness of the gel layer from 1 to 3 mm separately. The authors used supplement of aprotinin to control fibrin gel degradation and mechanical and chemical fixations together with microstructure culture dishes and poly-L-lysine tissue to control shrinking. The authors concluded that fibrin gel is found to be mouldable, and its degradation can be controlled by the use of aprotinin. Furthermore, a three-dimensional scaffold made from fibrin gel was used to develop a cell culture system of rapid tissue regeneration without any toxic degradation or inflammatory reactions (Ye et al., 2000). To achieve this, human aortic tissue was harvest from the ascending aorta, and pure human myofibroblasts were formulated and subsequently mixed with fibrinogen solution and left for four weeks to culture with various degrees of concentrations of aprotinin. The resultant 3D fibrin gel structures showed homogenous cell growth and confluent collagen production with no toxic degradation or inflammatory reactions. It was further stated that fibrin gel myofibroblasts structures liquefied within two days in medium without aprotinin, but medium added with higher concentration of aprotinin reserved the three-dimensional structure and had a maximum collagen content ($P < 0.005$) and an effective tissue production. However, it should be stated that although 3D fibrin gel structure has excellent seeding properties and controlled degradation, scaffold made of fibrin gel suffers from a number of limitations such as poor mechanical integrity, shrinkage and small initial mechanical stiffness, which certainly limit its potential for tissue-engineering applications.

Another material that has been tried for heart valve tissue engineering is chitosan, which, as discussed in Chapter 4, is a derivative of chitin, the polysaccharide found in crustaceans. This material has been previously proposed as a possible constructs for cartilage, and liver (Cuy et al., 2003; Suh and Matthew, 2000; Brody and Pandit, 2007) and has various advantages such as minimal rejection and foreign body reaction rates, manageable degradation, with similar physical structure to GAGs. Moreover, it was stated that the accessibility of hydroxyl and amino functional groups with it, which could be contained with the molecules, could be used to produce bioactive scaffolds (Suh and Matthew, 2000). Moreover, chitosan is known to encourage the growth of VECs and, is preferred over TCP, gelatin, poly-L-lactide-co-glycolide, and PHA respectively. However, findings from literature suggest that coating chitosan with proteins would not enhance VEC attachment; however, chitosan-collagen type IV films enriched VEC proliferation and morphology in comparison to that of chitosan only. Chitosan I snow considered to be an ideal material for various tissue engineering applications (Brody and Pandit, 2007).

In the study of Cuy et al., (2003), isolated aortic VEC cells were used, and it was found that VCE embrace especially to fibronectin, collagen types IV and I over laminin and osteopontin in a dose-dependent fashion. Moreover, a comparison study of the effect of the scaffold materials carried out by the authors showed VEC growth and morphology to be preferential in the following order: tissue culture polystyrene found to be greater than gelatin, poly (DL-lactide-co-glycolide) and chitosan greater than poly(hydroxy alkanoate). In addition, though the initial cell adhesion to protein-pre-coated chitosan was higher than for polystyrene, the attached protein pre-coating of chitosan did not meaningfully improve VEC growth, in spite of equivalent protein adsorption as to polystyrene (Cuy et al., 2003). Equally, as an alternative, protein pre-coatings chitosan/collagen type IV films constructs showed an enhancement of VEC growth and morphology over chitosan alone, implying that the use of chitosan improves polymer surface morphology for enhancement of cell adhesion and proliferation and differentiation.

In the study of Shah et al., (2008), the valvular interstitial cells (VICs) were condensed in a scaffold of cross-linked hydrogels produced from hyaluronic acid (HA) that was enzymatically degradable together with poly (ethylene glycol) (PEG). It was observed that the rate of degradation upon the introduction to bovine testes and release of HA fragments from the copolymer gels depended on the PEG content of the network. Moreover, these hydrogels showed permitted diffusion of ECM expanded by 3D cultured VICs and distributing the development of a specific matrix structure. In addition, it was found that initially the division of hydrogel cross-links before the network mass loss allowed the diffusion of collagen to occur, whereas at the later phases of degradation,there was stimulated elastin amplification and suppress collagen assembly due to HA fragment release (Shah et al., 2008). Table 7.3 give a list of tissue-Engineering invstigations using natural materials.

7.3.2. Polymer Scaffolds

Overview

It is recognised that scaffolds constructed from decellularization of xenograft are still considered to be immunogenically active and potentially can transmit the disease from donor to host (Schleicheret al., 2009; Geffre et al., 2009). Moreover, allografts constructs are very scarce and not available in large numbers for decellularization. Scaffolds constructed from biological materials, such as Fibrin, Collagen and Chitin based, are known to have insufficient mechanical integrity and high degree of shrinkage (Ye et al., 2000; Jockenhoevel et al., 2001b; Yang et al., 2001). Moreover, physical properties of biological materials fluctuate from batch to batch, and they are not suitable for mass production and are difficult to fabricate (Yang et al., 2001). Synthetic materials, on the other hand, can be tailor-made to suite various geometries and are easy to manufacture. Synthetic materials that have been approved by the FDA and have been examined by various researchers as possible scaffolds for various tissue-engineering applications are PLA, P4HB, PLLA, PHA, PVA and PGA copolymers (Bhattarai et al., 2005; Oliveira et al., 2006; Pham et al., 2006).

As stated in Chapter 4, polymer scaffold should be designed and fabricated to mimic as much as possible the physiological behaviour of the natural aortic valve. In case of aortic valve, the structure is very complex and consists of three leaflets that open and close during heart contraction and expansion and has three sinuses, cavities behind the leaflets. In design and constructing the valve, there are important challenging issues in mimicking the behaviour of the mechanical structures of the compliant aorta wall and the valve leaflets and their interaction with the blood flow. In a normal valve, the maximum mechanical stresses experienced by the valve occur at points of maximum flexion, which include the point of leaflet attachment and the line of coaptation (Morsi et al., 2004). Moreover, it is also recognised that during the cardiac cycle, the valve experienced significant stress variations and deformation of the leaflets, and the total stress on the valve becomes the sum of normal shear and bending stresses. At Swinburne, numerous evaluations of the haemodynamic of the various valve designs have been carried out both experimentally and numerically (Morsi and Birchall, 2005). Subsequently, during the design stage, it is important to consider features that will encourage the cell proliferation and differentiation, flow and transport of nutrients and wastes, and synthesis of extracellular matrix.

The most important physical properties of the developed polymer scaffold include porosity for cell ingrowth, a construct that surface that stabilities hydrophilicity and hydrophobicity factors to encourage adhesion of cells, mechanical integrity that are harmonious with those of the host tissue, and degradation rate as well as the by-product production. In addition to these possibilities, the scaffold can also be constructed to incorporate additives or active agents for enhancement of tissue generation (González et al., 2009; Yang et al., 2001). Researchers who adapted the polymer approach, including Shinoka et al. (1995), for example, have examined the possibility of engineering heart valve leaflets *in vivo* (lambs) by culturing a synthetic biodegradable scaffold manufactured from polyglycolic seeded with fibroblasts and endothelial cells. Two ovine arteries were used to harvest endothelial and fibroblasts cells, and these cells were subsequently used to seed the fibre scaffold, first by fibroblasts and then by endothelial cells.

Table 7.3. Tissue-Engineering Studies using natural materials

Authors	Objectives and the type of scaffold used	Cell seeding and techniques	General remarks
Vesely et al. (1998)	Establishing the link between elastin and collagen structures .Valve constructed from type I collagen bovine skin source.	The valve was seeded with porcine heart valve, myofibroblasts &- porcine aortic endothelial cells.	Collagen is suitable for HVTE Sufficient amount of myofibroblasts Expression of ECM Generation of endothelial cell layer was evident
Rothenburger et al. (2002)	Investigate the use of type 1 Collagen for tissue engineering of heart valve.	Myofibroblasts & endothelial cells were used. Histological, immuno histochemical and cupromeronic blue staining methods.	Proteoglycan heart valve tissue exhibits similar behaviour as natural heart
Taylor et al. (2002)	Optimise the use of collagen sponge as scaffold for TE.	Interstitial cells (ICs)ICs cultured from cardiac valves and planted onto glass cover-lips/seeded in collagen sponge and then stained by immunofluorescence or immuno peroxidase	The findings showed that ICs were present in the native valve leaflet, particularly on the ventricular Cells were viable in the sponge after four weeks and also able to proliferate .The collagen sponge is a suitable biodegradable
Ramamurthi et al. (2005)	To examine the use of Non-biodegradable Hydrogels(hylans) containing cross-linked hyaluronan and glycosaminoglycan was used with the aim of examining the extracellular matrix (collagen and elastin)	Cultured neonatal rat aortic smooth muscle for six weeks. the hylan gels were surfaced-texturized by irradiation with UV light under hydrated conditions and then seeded with Neonatal rat aortic smooth muscle cells (NRASMCs)	A unique type of matrix layer, which is rich in elastin, was produced. The amount of elastin observed was greater by 25% for hylan-based cells. Cells cultured on hylan gels produced 56% less collagen of DNA than cells cultured on plastic
Shi et al. (2006)	Collagenous structure, casting into rectangular and branched well	The scaffold was seed with neonatal rat aortic smooth muscle and condition *in vitro* for six to eight weeks.	Multi-branched collagen structure has the weakest mechanical strength while long and linear structure has the strongest.
Neidert et al. (2006)	Used type 1 Collagen. Examining tissue equivalent method of entrapping cells in a biopolymer gel	Neonatal human dermal fibroblasts conditioned *in vitro* after six weeks	Similar collagen fibre alignments and geometry with native valve Neonatal human dermal fibroblasts (NHDFs) for the tissue engineered were present but lack of other ECM components.
Jockenhoevel et al. (2001b)	Fibrin gel scaffold using injection moulding	Patient's blood	Histological findings exhibited excellent tissue development with viable fibroblasts surrounded by collagen bundles with excellent ECM. Possible formation of complex structures is possible and degradation and polymerization can be controlled.

Authors	Objectives and the type of scaffold used	Cell seeding and techniques	General remarks
Ye et al. (2000)	Examining the use of fibrin gel for TE.	Human myofibroblasts conditioned statically in four weeks	Homogenous cell growth of fibrin gel structure, confluent collagen production with no toxic degradation or inflammatory reactions. Controlled degradation. Poor mechanical property with small initial mechanical stiffness and some shrinkage of the scaffold
Cuy et al. (2003)	Examining the effect of Scaffold materials on tissue generation, An alternative, composite chitosan/collagen type IV films with four biodegradable polymer substrates were examined.	Bovine aortic valve endothelial cell (VEC) VEC morphology at six days in culture varied a bit	VEC growth on chitosan occurred over a period of seven days. Chitosan added with appropriate protein, posses promising substrate for valve TE
Shah et al. (2008)	To examine the use of HA for TE. Photoinitiated chain copolymerization of HA-based and PEG-based macromers for gels of varying composition and hyaluronidase (HAase) mediated network degradation was used for the gel chemistry	Valvular interstitial cells (VICs) at concentrations greater than 1.38–5.50 μm These hydrogels demonstrate dual function of permitting the diffusion of ECM elaborated by 3D cultured VICs and development of specific matrix	HA was found to sustain VICs endurance for 14 days. Later stages of degradation creates elastin elaboration and suppress collagen production due to HA fragment release of the hydrogels The dual utility of using HA (structural and bioactive) component to direct VIC secretory functions is promising as a material for heart valve tissue engineering strategies but the ECM composition must be finely tuned

The fibroblasts developed into a tissue-like sheet, and endothelial cells formed a cellular monolayer encapsulating the leaflet of the valve. Subsequently, the tissue-engineered leaflets were implanted in seven animals, and in each animal the right posterior leaflet of the pulmonary valve was resected and substituted with the seeded fibre scaffold (Shinoka et al., 1995). The seven animals survived the operation, and it was reported that leaflet tissue engineered from its cellular components could function satisfactorily in pulmonary valve position, and a tissue-engineered valve leaflet is conceivable. The authors further suggested that the autograft tissue would probably be superior to allogenic tissue.

Since cells present in the heart valve leaflets are mainly fibroblasts, myofibroblasts and endothelial cells, it is advantages to utilize these cells for heart valve tissue engineering. Shinoka et al., (1995) isolated fibroblasts and endothelial cells from ovine arteries and used them for pulmonary valve leaflets in lambs

Zund et al., (1997) examined the creation of *in vitro* tissue-engineered heart valve employing cardiovascular cells on degradable polymer scaffolds and fabricated 40 heart valve leaflets from two sources. One source was the xenograft leaflets seeded with human dermal fibroblasts and bovine aortic endothelial cells, the second source was allograft valve leaflets seeded with a mixed sheep cell population of endothelial cells and myofibroblasts. The authors carried out a histological evaluation of these constructs and found that was no evidence of capillary formation from endothelial cells invading. However, they found that the myofibroblasts and smooth muscle matrix as well as the endothelial linings to be completed. The authors concluded that it is possible to regenerate allogenic heart valve tissue for valve construction.

Later on, Sodian's group examined the potential use of porous PHA and PHO as a TEHV scaffolds in various *in vitro* and *in vivo* trials (Sodian et al., 2000 (a); Sodian et al., 2000 (b)). Sodian et al., (2000a) attempted to create a tri-leaflets tissue-engineered heart valve using polyhydroxyalkanoate BD polymer with different degree of porosity (100 to 240 μm) by *in vitro* conditioning. The scaffold was seeded with vascular cells grown from ovine carotid artery and tested for one, four, and eight days, respectively. Extracellular matrix (ECM) formation was observed, and cell proliferation (DNA assay) and the capacity to generate collagen were noted. The same group reported their *in vivo* experience by the implementation of a whole tri-leaflet tissue-engineered heart valve in pulmonary position of a lamb model using the same type of scaffold i.e., Polyhydroxyalkanoate with pore size varied from 180 to 240 μm and the same vascular cells harvested from ovine carotid arteries. *In vivo* trials were 1, 5, 13, and 17 weeks, respectively, and all animal models survived the procedure. The histological analysis indicated that laminated fibrous tissue with predominant glycosaminoglycans as Extra Cellular Matrix (ECM). The authors reported that with *in vitro* experiments, the cells formed confluent layers on the surface of the material and grew into pores, whereas, *in vivo*, a confluent endothelial cell lining was not achieved. Moreover, although the Extra Cellular Matrix (ECM) was found to be rich in collagen and GAGs, elastin was not expressed.

Another significant study was carried out by Hoerstrup et al. (2000), in which the authors constructed a tri-leaflet heart valve porous scaffold from a flexible, thermoplastic biopolyester, a polyhydroxyoctanoate (PHO) and implanted into the pulmonary artery of the same animal model from where the cells were harvested in the first place. Again, it was reported that *in vitro* conditioning was capable of producing autologous tissue-engineered valve that could function up to five months and resembled normal heart valves in microstructure, mechanical properties, and extracellular matrix formation. (Hoerstrup et al., 2000).

Kadner et al. (2002), Rabkin et al. (2002) and Hoerstrup et al. (2000) studied the possibility of coating PGA in P4HB to improve the mechanical performance of the PGA. The findings indicated that a comparable uni-axial reaction to that of the native pulmonary valve leaflet can be achieved but with low degree of flexibility and deformation. The various breakages and fragmentation of the scaffold material *in vitro* question the appropriateness of this scaffold *in vivo*.

Nuttelman et al. (2002) used Poly(Vinyl alcohol) (PVA) with grafted PLA side chains to develop degradable and photo-cross-linkable polylactic acid-g-PVA for scaffold material. The scaffold was seeded with Porcine Aortic VICs for one day *in vitro*. In their study, the Nutellman group showed that this material had a great potential for tissue engineering as the rate of degradation, mass erosion profile and hydrophobicity could be controlled. Their experiment also showed that valve interstitial cells (VICs) attachments were evident.

Masters et al. (2004) carried out another study for designing scaffolds for interstitial cells (VICs) and used Poly(ethylene glycol) PEG containing RGD and PHSRN (a fibronectin derived peptide sequence) as scaffold material. PEG is a non-cell bond material, which offers an ideal surface to examine the effect of peptide modification on cell adhesion. Subsequently, they seeded the scaffold with the porcine aortic VICs and conditioned it for 72hrs in static culture. The authors found out that the use of RGD and PHSRN only slightly improved VICs proliferation, and adhesion of these cells were not adequate.

Moreover, on the effect of chitosan coating, Mei et al. (2005) used a scaffold made of a biocompatible of poly(e-caprolactone) PCL that was coated with different concentrations of chitosan from (0.5 to 2%) and seeded with human fetal lung fibroblasts and conditioned for one week *in vitro*. The findings indicated that chitosan coating enhanced cellular adhesion and proliferation, while its adsorption continued to be stable throughout the testing, and PCL scaffold morphology was partially influenced by chitosan coating.

In the same year, Zong et al. (2005) examined the effect of structural and functional features of scaffolds of various micro and sub-micron morphological structures on the growth and differentiation of cardiac myocytes. To obtain anisotropy fibre orientation the authors used post-processing in electrospinning manufacturing technique to fabricate a biodegradable non-woven poly (lactide)- and poly (glycolide)-based (PLGA) scaffolds. Moreover, a dose-response influence of the poly (glycolide) on the degradation rates and pH value changes was assessed using *in vitro* experimentation. Scanning electron microscopy (SEM) analysis revealed that the fine fibre architecture of the non-woven matrix assisted the cardiomyocytes to effectively improve the external signals for isotropic and anisotropic growth and increase interconnectivity of the fibres within the scaffold. Moreover, the results of confocal microscopy showed that cardiomyocytes preferred hydrophobic surfaces structure. It was stated that the chemistry and geometry of the constructed nano- and micro-textured surfaces of the scaffold could regulate the physical structure and the effectiveness of the engineered cardiac tissue. This research in addition to others showed the flexibility of electrospinning manufacturing techniques to fabricate micro-and nano-biomaterials scaffolds for tissue-engineering applications with excellent morphology for cell adhesion, differentiation and growth (Zong et al., 2005).

In the same year, scaffold made of a composition of non-woven PGA and PLLA with proportion of 50% of each and seeded with ovine carotid artery smooth muscle cells was *in vitro* conditioned for three weeks, and the results of seeded and unseeded scaffolds were analysed by Engelmayr et al., (2005). Subsequently, the effective stiffness (E) of the scaffolds was examined under static condition (static group) and unidirectional cyclic three-point flexure condition in a bioreactor (flex group). The findings indicated that the scaffolds in flex group had decreased cellular necrosis, with no elastin and an increased stiffness, E by 429%, as well as cellular infiltration and collagen expression (Engelmayr et al., 2005). It is interesting to note that scaffold that was seeded with ovine bone marrow (mesenchymal stem cells MSCs) showed incomplete coaptation and mild regurgitation, however, fibrosa, spongiosa and ventricular layers of similar characteristics of native heart valve were noted.

Lieshout et al. (2006), used electrospinning and knitted techniques to construct two different scaffolds for tissue engineering of the aortic valve (Figure 7.3). These scaffolds were conditioned in a physiologic flow, seeded with fibrin gel enclosed human myofibroblasts and cultured for 23 days under continuous medium perfusion condition. It was observed that when subjected to physiological hemodynamics conditions the spun scaffold structure totally damaged within six hours, whereas the knitted scaffold lingered unharmed. The authors evaluated the tissue formation with the aid of confocal laser scanning microscopy, histology and DNA quantification. Moreover, hydroxyproline assay was used to examine collagen formation, and it was noted that cell adhesion and proliferations were good for both types of scaffolds. However, cellular infiltration into the spun scaffold was unsatisfactory. Nevertheless, the regeneration of collagen normalized to DNA quantity was almost the same for both scaffolds, but seeding efficiency was reported to be higher for the spun scaffold. In addition, the knitted tissue scaffold demonstrated complete cellular penetration into the pores with a good interconnectivity. It was then suggested that an optimal scaffold could be a combination of the strength of the knitted structure and the cell-filtering ability of the spun structure.

Courtney et al. (2006) found that these mechanical characteristics were not possible with current polyglycol acid/polylactic acid (PGA/PLLA) materials commonly used in tissue-engineering applications. They fabricated electrospun poly (ester urethane) ureas and modeled a range of electrospun polyurethane in an effort to quantify the effect of fibre orientation on the mechanical properties of the scaffold. It was reported that since electrospun polyurethane scaffolds are elastomeric, they could be constructed with the correct morphology of the surface, with well-aligned and interconnected fibre networks. Such type of scaffolds would provide a good mechanical integrity and assist in quality of the tissue addition and formation (Courtney et al., 2006).

More recently, copolymers of non-woven PGA and PLLA have been used to construct a valve and tested *in vitro* as discussed *by* Brody and Pandit (2007). The scaffolds were seeded with ovine bone marrow MSCs (Mesenchymal Stem Cell) and implanted into an autologous model. The findings indicated although, incomplete coaptation and mild regurgitation were reported, possibly due to cusp thickness, the constructed valve functioned satisfactory for four months. However, particular significance of this study, is the site-appropriate phenotype facilated by the MSCs; morphological remodelling of the cells and the subsequent three distinct layers that were created in the construct. These layers were equivalent to the fibrosa, spongiosa, and ventricular ones of the native heart valve, but elastin again was not noted (Brody and Pandit, 2007).

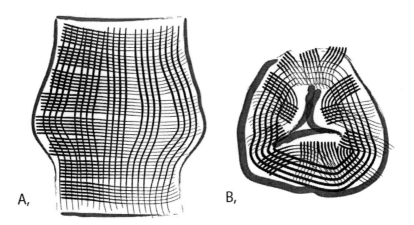

Figure 7.3. Illustration of the knitted valve (left) on top of the stainless steel mold that forms the sinuses and (right) after application of three fibrin gel layers and removal from the mold.

Schaefermeier et al. (2009) carried out a study to design and fabricate a 3D scaffold for TEHV. The authors used Poly-4-hydroxybutyrate (P4HB) for scaffold material and then fabricated the scaffold by the following steps: Firstly, the computed tomography (CT) scan was obtained from an aortic homograft, and the image was processed by Amira-Anaplast software to retrieve the image segmentations and then reprocessed by region-growing technique. Then, collected 2D segmented images were connected together to become a 3D image of the aortic valve. Established 3D image was converted to stereo-lithographic (STL) model and fed into FDM3000 stereo-lithography machine. The machine fabricated resinic STL valve model and negative cast of ventricular side of prototype. Constructed cast was used to form the P4HB scaffold by pressing the polymer into the cast and thermal processing technique. The authors tested the scaffold in pulsatile flow bioreactor and synchronous opening and closure under various hemodynamic conditions. However, mild stenosis and regurgitation were noticed.

The *in vitro* bioreactor study of Schaefermeire et al. (2009), who fabricated a 3D scaffold made of Poly-4-hydroxybutyrate (P4HB) and synchronous opening and closure under normal and critical physiological flow and pressure conditions, reported mild stenosis and regurgitation. Another interesting study was carried out by Yamanami et al. (2010), who constructed a complete first type of bio-valve with sinus of valsalva. Two molds were designed for the convex and concave shapes to construct the sinus of valsalva, and the leaflets of the valve were made from silicon. The authors also covered the scaffold with porous polyurethane mesh. Cell proliferation and development were carried out in the dorsal subcutaneous pouch of the rabbit, and after one month the valve was harvested and conditioned *in vitro* using pulsatile flow circuit model. It was observed that the valve opened and closed synchronously without hitting or flapping against the scaffold wall, and 20% regurgitation under 170mL/min condition was found. Although this study showed the new approach to construct the valve with sinus of valsalva, and cells proliferation *in vivo*, the moderate regurgitation is still considered to be an issue of importance. Moreover, a wider range of haemodynamic forces need to be carried out to confirm the structure integrity of the valve developed by the authors.

Table 7.4 illustrates a summary of the development of tissue engineering heart valve using polymers approach.

7.4. Source of Cells

It would be best if the ideal cell source for required application could be determined on a cost effective way. With tissue engineering of heart valve, VCs are mainly fibroblasts, myofibroblasts and endothelial cells, and various researchers have used these cells harvested from animals and humans for heart valve tissue engineering. For example, Shinoka et al., (1995) isolated fibroblasts and endothelial cells from ovine arteries and used them for pulmonary valve leaflets in lambs, whereas, in the study of Curtil et al., (1997), porous scaffold was seeded with low-density human dermal fibroblast and umbilical endothelial cells. Similarly, Hoerstrup et al., (2000) and Steinhoff et al., (2000) made use of autologous ovine arterial myofibroblasts from lamb and then fixed endothelial cells on the top of them. Subsequently, various other researchers followed this approach using autologous myofibroblasts and layering them with endothelial cells (Ye et al., 2000; Jockenhoevel et al., 2001 (a, b)). In another study, Zeltinger et al., (2001) seeded decellularized aortic porcine heart valve with human neonatal dermal fibroblasts (DmFbs). Later, Dohmen et al. (2003) seeded ovine autologous vascular endothelial cells (AVECs) onto decellularized porcine pulmonary. The porcine aortic valve decellularized by trypsin/EDTA detergent was seeded with ovine myofibroblasts and subsequently coated with endothelial cells (Leyh et al., 2003). Later, Lichtenberg et al., (2006) decellularized ovine pulmonary valve and subsequently seeded it with endothelial cells (ECs). Recently, Hong et al., (2009) seeded mesenchymal stem cells (MSCs) on decellularized porcine aortic heart valve scaffold.

However, it can be argued that since the ultimate objective of these studies was to generate tissue-engineered heart valves for human usage, cells need to be from human origin. The most suitable cells for this purpose would be vascular-derived cells, cells obtained from the bone marrow, umbilical cord, blood and even chorionic villi (Schmidt and Hoerstrup, 2005). Hoerstrup et al., (1998) isolated myofibroblasts and endothelial cells from human peripheral arteries, and pure cell lines were developed using cell sorter. Myofibroblasts were seeded onto a biodegradable synthetic biomaterial and then layered with endothelial cells. Zund et al. (1998) obtained myofibroblasts from human foreskin and seeded onto resorbable PGA mesh and after three weeks of culturing, seeded endothelial cells were obtained from human aorta. Hence, Schnell et al., (2001) compared myofibroblasts from vascular origins. Myofibroblasts obtained from ascending aorta and saphenous vein were cultured on biodegradable polyurethane biomaterials and were studied for cell viability and extracellular matrix production.

Table 7.4. Tissue engineering of heart valve with Polymers

Investigators	Scaffold Material and fabrications	Cells source seeding observation	Comments on mechanical properties ECM generation and degradation of scaffold
Shinoka et al. (1995)	PGA woven mesh surrounded by two non-woven PGA mesh sheets	Ovine Endothelial cells and fibroblasts, three autologous and four allogenic ovine valve condition *in vivo* for 21 days	Autologous valve: no stenosis and regurgitation, size and shape remain intact. Allogenic valve: similar tensile strength before implantation, less flexible and thicker cusp than native valve Collagen content on fabricated valve had 30% less than native valve
Zund et al. (1997)	Composite of PGA with copolymer of PGA and PLAA Non-woven of PGA fibres (for outer layers of leaflets) and a woven mesh of PGA copolymer layer (for inner layers of leaflets) thermally fused together.	Human fibroblasts for xenograft leaflet and bovine endothelial cells for allograft leaflet condition *In vitro*, for 21 days	Fibroblasts attached into scaffold and had begun to spread out Poor mechanical properties reported No comment on degradation of scaffold
Sodian et al. (2000) (a,b,c and d)	PHA Stent fabricated by wrapping PHA around heart valve cast, leaflet were welded outside of stent by thermal technique then scaffold pressed into aluminium cast.	Ovine Vascular cells and the scaffold was *in vitro* condition for four days	Synchronous opening and closure achieved under pulsatile flow bioreactor Observation of cells attachments grow into were made
Hoerstrup et al. (2000)	PGA coated with thin layer of P4HB Heat application welding technique	Autologous ovine myofibroblast and endothelial cells, *In vivo* implementation for 20 weeks	Similar mechanical characteristics to native valve Good cell proliferation on the TE leaflets, Elastin was detectable in the TE leaflets by six weeks The scaffold totally degraded after eight weeks
Kandner et al. (2002)	No-woven mesh PGA coated with P4HB	Human umbilical cord cells, *In vitro* conditioning for three weeks.	Extracellular matrix analysis demonstrated deposition of collagen I and III, In contrast, the tissue-engineered patches reached only 34% of the amount of glycoaminoglycans measured in native pulmonary artery
Investigators	Scaffold Material and fabrications	Cells source seeding observation	Comments on mechanical properties ECM generation and degradation of scaffold
			Degradation of the scaffold was studied by multiple breakages and fragmentation of the polymer fibres
Nuttelma et al. (2002)	PVA based hydrogel with PLA side chain	Porcine aortic valve interstitial cells (VICs) *in vitro*, 72 hours	Great ability to control the rate of network degradation, mass erosion profile Cells attachment to the scaffold were observed Degradation rate of PVA depends on relative length of kinetic chain. (controllable)

Table 7.4. (Continued)

Investigators	Scaffold Material and fabrications	Cells source seeding observation	Comments on mechanical properties ECM generation and degradation of scaffold
Hoerstrup et al. (2002)	Non-woven mesh PGA coated with thin layer of P4HB. Heat application welding technique	Human marrow stromal cells *In vitro*, seven days in static nutrient culture followed by 14 days in pule duplicator system	Synchronous opening and closure achieved. Similar pliable mechanical properties to native valve. In high fellow and pressure bioreactor all leaflet remain intact. Immunohistochemistry showed positive staining for collagen types I, III, αSMA, and vimentin. desmin, collagen types II and IV were not detected Multiple breakage and fragmentation of polymer fibres were observed
Perry et al. (2003)	Two sheets, of non-woven PGA mesh dip coated in 1% P4HB Moulded to form the valve	Ovine bone marrow cells	Cusps coapted uniformly in close position. Cells grew in confluence and conduit wall observed. Mechanical properties that resemble that of native valve leaflet tissue. Cells evenly distributed tissue over the entire porous scaffold Degradation rate improved by coating PGA in P4Hb so the tissue is able to form before the polymer loses its mechanical strength
Master et al. (2004)	Poly(ethylene glycol) PEG modified with fibronectin-derived peptide sequences	Porcine aortic VIC condition *In vitro*, up to 72 hours in static nutrient cultures	Proliferation improved slightly by RGD and PHSRN, Poor cells attachment into the scaffold observed No comments on the rate of degradation.
Engelmayr et al. (2005)	Non-woven 50:50 blend of PGA and PLLA fibres Scaffold samples were cut to (25 x 7.5 x 1 mm)	Ovine carotid artery smooth muscle cells	Effective stiffness (E) evaluated for scaffolds in two groups. Scaffolds in flex group had increased E by 429% Flexure increased collagen content and cellular infiltration and decreased cellular necrosis
Mei et al. (2005)	PCL coated with different concentration of chitosan from (0.5 to 2%) Particle leaching/ casting	Human fetal lung fibroblasts	PCL scaffold morphology slightly affected by coating of different concentration of chitosan Cells attachment and to the scaffold and cellular proliferation greatly improved by chitosan coating

Investigators	Scaffold Material and fabrications	Cells source seeding observation	Comments on mechanical properties ECM generation and degradation of scaffold
Zong et al.(2005),	Poly(glycolide-co-lactide) (PLA10GA90, LA/GA 10:90) random Electrospinning technique	Primary Cardiomyocytes (CMs) of Sprague-Dawley Rat	Scaffold is pliable and provide enough guidance for CMs growth Cells attachment depends on electrospun PLGA composition. The hydrophilic surface of PLGA and PEG-PLA has adverse effect of cell proliferation. CMs cultured on (PLGA10GA90 and PLLA) lose spatial organization and clustered due to fast degradation rate Degradation rate is controllable
Schaefermeire, (2009)	P4HB Stereo lithography	Not reported	Valve open and close synchronously with mild stenosis and regurgitation
Chen et al. (2009)	PCL for the ring, Thermoplastic Polyurethane TPU for Valve leaflets FDM for the ring, Electrospinning for Valve leaflets	Not used	Demonstrated that mechanical properties of the scaffold such as pore size and stress-strain curve of scaffold can be adjusted
Yamanami et al. (2010)	Polyurethane Molding techniques	Connective tissue membrane from rabbit. Condition *In vivo* in one month	Valve open and close synchronously, little stenosis reported in valve root, and regurgitation prevented in diastole phase.

Since the cells obtained from other human donors might cause an immune response in the patient's body, use of stem cells was considerd—especially MSCs since embryonic stem cells are hard to control. MSCs can be obtained from its sources in a non-invasive manner for heart valve tissue engineering. Perry et al., (2003) used ovine MSCs isolated from bone marrow of sheep and cultured on a synthetic biomaterial. Sutherland et al., (2005) went further to use autologous ovine MSCs to seed a synthetic biomaterial (in shape of valve), which was implanted in pulmonary valve position of sheep. Hong et al., (2009) used rat MSCs to seed decellularized porcine heart valves. Iop et al., (2009) made use of human bone marrow MSCs on decellularized porcine pulmonary valve leaflets. Depending upon the scaffold materials selected and culture technique employed, most of the above mentioned studies reported positive results and experience using MSCs as source of cells.

Moreover, in case of decellularized valves, a layer of endothelial cells might prevent thrombus formation, and hence this possibility was explored. A variety of studies have been carried out making use of endothelial progenitor cells (EPCs)—these are the cells that have the ability to form endothelial cells. Since EPCs are not completely differentiated cells, they do not evoke an immune response similar to endothelial cells and are thereby preferred. Kaushal et al., (2001) used EPCs obtained from sheep's peripheral blood on decellularized porcine vessels. Schmidt et al., (2004) used human umbilical cord vein derived EPCs on polymeric biomaterials. Fang et al., (2007) used umbilical cord blood derived EPCs onto decellularized porcine aortic valves. Thus, EPCs were mainly utilized to make the top-most endothelial layer in tissue-engineered constructs.

7.5. Summary of Findings and Discussion

In this chapter, the progress of developing tissue engineering heart valve TEHV using natural and polymer scaffolds have been introduced. With the *approach of decellularization,* the donor heart valves or animal-derived valves are less immunogenic due to the depletion of cellular antigens, and the approach has been used by a number of researchers (Mol, 2005). The rationale for such an approach is based on the fact that decellularized scaffolds inherent cell compatibility and excellent topography and architecture coupled with good functionality and mechanical integrities that are almost identical to that of nature valve. Moreover, the depleting of cellular components are generated inside a material, which composed of extracellular matrix proteins that forms the basic construct for cell differentiation and growth (Prabha and Verghese, 2008). Moreover, various decellularization techniques have been used that will remove cells and do minimal or no harm to the ECM – physical, chemical and enzymatic method (Gilbert et al., 2006). The possibility of decellularization is most effective by using these three methods in combination with each other. Initially, cell membrane would be lysed using any of the physical methods (like sonication, mechanical pressure, freezing or thawing) or by the use of ionic solutions, after which the cellular components are separated from the ECM by enzymatic treatments. The cytoplasmic and nuclear cellular components are solubilized using detergents, and then cellular remnants are removed from the tissue in the end. It has been noted that the effectiveness of decellularization technique is dependent upon the tissue in question, as one decellularization method could be successful for one type of tissue but might not work for the other.

As stated above, this approach has been utilized by previous researchers using different decellularization techniques; for example, Curtil et al. (Curtil et al., 1997) has chosen freeze-drying technique, whereas Kasimir et al. (2005), Leyh et al. (2003) and Steinhoff et al. (2000) elected to use trypsin/EDTA. In general, however, detergent treatment method is considered to be an effective technique and could give a total removal of cells as well as preservation of the complete matrix of the construct (Bader et al., 1998; Bertipaglia et al., 2003; Leyh et al., 2003; Kasimir et al., 2005), multi-step enzymatic procedures (Zeltinger et al., 2001). Still, it has been argued by Booth et al. (2002) that the discrepancy observed using different decellularization methods could be ascribed to proteases present in the native tissue of the valve. The activation of them can lead to autolysis (destruction) of extracellular matrix proteins and, as a result, damage in the structure and function of the cell matrix. Consequently, it was suggested that protease inhibitors must be used with decellularization process (Booth et al., 2002).

As observed, decellularized valves are not ideal biomaterials. With respect to the use of natural scaffolds, it has been argued that utilizing biological materials of similar physical properties of natural valve tissues can assist in maintaining the natural environments that enhance physiological events such as cell attachment (Lee et al., 2011). However, biological materials do not offer the flexibility of custom-designed properties such as specific surface structure or pore size, which affects both the mechanical and cellular characteristics of biomaterials.

Collagen is one of the biological materials that demonstrates biodegradable characteristics, which are significant factors in tissue engineering, and has been used as a foam (Rothenburger et al., 2002), gel or sheet (Hutmacher, 2001), sponge (Taylor et al., 2002), and even as a fibre-based scaffold (Rothenburger et al., 2001). Nevertheless, human collagen is scarcely available, and hence, only animal collagens are used. More importantly, the slowness of degradation could lead to residual of collagen remaining after *in vivo* grafting with serious adverse effects such as the risk of infectious disease (zoonosis) as well as immunological reactions and inflammation.

Another biological material discussed here in this chapter is fibrin, which can be obtained from patient's blood and hence is considered to be autologous without any toxic degradation or inflammatory reactions. Fibrin has excellent and well-regulated biodegradable properties, which can be restrained by injection aprotonin, a proteinase inhibitor to slow or sometime even halt the process of fibrinolysis (Lee et al., 2001). Moreover, with the aid of moulding of the cell-gel mixture, tracked by enzymatic polymerization of fibrinogen, it is possible to construct a 3D scaffold and control growth factors in specific areas (Ye et al., 2000 (b), Schense and Hubbell, 1998).

However, fibrin scaffold has a low level of diffusion and washout of substances (Jockenhoevel et al., 2001 (b)) and is inclined to shrink rapidly with limited mechanical properties (Jockenhoevel et al., 2001 (a)). However, using poly-L-lysine fixation with the gel can considerably improve mechanical and shrinkage properties. Still, fibrin gel could also be utilized as a cell carrier in porous synthetic scaffolds (Ameer et al., 2002).

Chitosan obtained from crustaceans has good biocompatibility and has structural similar to GAGs and therefore has hydroxyl and amino functional groups (Suh and Matthew, 2000). Moreover, Chitosan is also known to assists in the growth of vascular endothelial cells (VECs), when compared to TCP, gelatin, poly-L-lactide-co-glycolide, and PHA. However, in relation to optimization of VEC attachment, proliferation, and morphology, chitosan is thought to be as effective as, or even better than, all other constructs, except TCP (Brody and Pandit, 2007). Cuy et al., (2003) seeded VECs on chitosan biomaterial, which showed good cell proliferation. However, chitosan coupled with collagen type I biomaterial showed improved biocompatibility.

Hyaluronic acid (HA) happens to be an element of cardiac jelly and thus has an influence on valvular cellular behaviour (Shah et al., 2008). This biomaterial is normally used as a hydrogel and when seeded with VICs, it resulted in matrix synthesis with good cell proliferation throughout the cells (Ramamurthy and Vesely, 2005). Cellular attachment on these biomaterials can be intensified by cross-linking them by the aid of ultraviolet light (Ramamurthy and Vesely, 2005). Similar to the above-mentioned naturally available biomaterials, even HA hydrogels have low mechanical properties and hence cannot bear the high pressures at aortic position.

However, since the biological biomaterials cannot fulfil the conditions required for the role as biomaterials for tissue-engineered aortic valve, synthetic biomaterials need to be examined for this purpose. With the concept of synthetic scaffolds, different types of materials have been used as scaffold for various types of tissue engineering (Lee and Mooney, 2001).

Moreover, in polymer scaffold approach, the constructs generally are well optimized and can be easily constructed with specific properties in mind. Potentially various biocompatible, resorbable, materials with well-defined macrostructure and interconnectivity that facilitate in-growth, nutrient supply and waste elimination could be manufactured. This bioresorbable scaffold is used as a temporary matrix until the seeded cells are capable of producing their own ECM. Moreover, the rate of degradation is generally tailored to the application and the rate of new tissue regeneration (Hollister, 2005; Karageorgiou and Kaplan, 2005; Liu and Ma, 2004; Mendelson and Schoen, 2006).

This approach, as stated above, proved to be popular, and research over the past two decades has provided a wealth of knowledge regarding synthetic biodegradable polymers for scaffold fabrication. These materials include polylactic acid (PLA), polyglycolic acid (PGA), copolymers of PLA and PGA (PLGA), polycaprolactone (PCL), polyanhydrides, poly(ortho esters), polycarbonates and polyfumarates. However, it should be pointed out that no single material can fulfill the required characteristics for the fabrication of the correct constructs in terms of mold-ability mechanical integrity and rate of degradation as well as the inflammatory response. As discussed in Chapter 5, different techniques can be used to create a scaffold of the required characteristics, including woven or non-woven fibre mesh or using the simple construction fabrication technique such as a salt-leaching technique (Agrawal and Ray, 2001; Hutmacher, 2001). Moreover, when it comes to the construction of heart valve leaflets, special requirements of the materials to be used must be satisfied. These requirements must ensure that the flexibility and deformation of leaflets are similar to those of biological ones. In literature, attempts were made to construct leaflets of heart valve from various combinations of aliphatic polyesters; Shinoka's group used woven polyglactin and non-woven PGA meshes (Shinoka et al., 1995, 1996; Shinoka, 2002) whereas others used films of PLGA and non-woven PGA meshes (Zund et al., 1997; Kim et al., 2001).

Alternatively Sodian's group capitalized on the fact that polyhydroxyoctanoate (PHO) and poly-4-hydroxybutyrate (P4HB) are thermoplastic materials, and, therefore, they can be used in, for example, Rapid prototyping, Fused Deposition Modelling where the exact shape of the valve conduits including and sina of valva can be easily and accurately produced

(Sodian et al.,, 2000a,b,c,d). Later, Hoerstrup's group proposed the use of combinations of P4HB and PGA and reported good results (Hoerstrup et al., 2000; Rabkin et al., 2002). Nevertheless, there is still a challenge with the use of aliphatic polyester to balance between the rate of degradation and required flexibility of the valve, particularly the leaflets. For aortic valve, the deformation of the aorta is around 15% during heartbeat, which imposes still a huge task to develop materials that are amenable to this physiological condition.

Schnell et al. (2001) examined the influence of the source of vascular-derived cells on the regeneration of the tissue and found that, for example, scaffold seeded with venous cells displayed better characteristics than the ones seeded with arterial cells in terms of the percentage of collagen generation and mechanical properties (Schnell et al., 2001). On the influence of the source of endothelial cells, Sodian et al. (2002) found no endothelial lining of the construct when these cells were obtained from a vein but still are not clear why there is a difference between endothelial cells from arteries and veins and what the exact effect of *in vitro* on the tissue formation and coverage of endothelia cells is.

As stated above, in the work of Zong et al. (2005), elctrospinning technique was used to construct three scaffolds, made from PLLA, PLA75GA25+PEG-PLA and PLA10GA90+PLLA, with an averged fibre diameter of 1µm approximetly to determine the effect of these type of materials on heart tissue regeneration (Zong et al., 2005). Comparable surface morphology and degree of porosity in the range of 71% for PLLA to 78% for PLA75GA25+PEG-PLA were achieved. It was reported that hydrophilic surface of PLGA and PEG-PLA had an adverse effect on cell proliferation, producing clusters that compromised cell differentiation, and the most hydrophobic scaffold out of them was PLLA, which showed the best support for CMs attachment and structural development. Moreover, it was also reported that fast degradation rate of scaffold negatively affected the cardiomyocytes by losing spatial organization and cluster together. It should be noted that the range of porosity used in this study was less than desireable for tissue engineering of heart valve (95% and more).

Moreover, various authors have used non-Woven PGA dip coated with P4HB as scaffold for heart valve with different types of cells (Kadner et al., 2002; Hoerstrup et al., 2002; Dvorin et al., 2003; Perry et al., 2003; Mol, 2005). Perry,s group investigated bone marrow as a cell source for tissue engineering heart valve. The scanning electron microscopy (SEM) and biomechanical flexure analysis showed that tissue generated over the cusps was uniformly disturbed and had a mechanical stiffness similar to the natural valve, concluding that although the cell distribution and tissue generated over the leaflets were considered to be acceptable, the scaffold material did not degrade, and even after three weeks of conditioning residual polymer fibres were evident. Kandner's group, on the other hand, seeded the scaffold with human umbilical cord cells (UCC) and *in vitro* conditioned the scaffold for, again, three weeks. They evaluated the constructed valve by uniaxial stress testing and found that uniaxial stress responses were similar to those of the native pulmonary valve. However, the regenerated tissue with the scaffold was less flexible and distensible. The one-week *in vitro* bioreactor testing of Hoerstrup's group, using human marrow stromal cells, showed that the resultant TEHV was effectively synchronous, opening and closing with viable ECM formation and tissue. Also, authors claimed that mechanical properties of regenerated valve leaflets were comparable to native ones. Dvorin's group, on the other hand, seeded the scaffold with ovine aortic cells (VECs) and circulating endothelial progenitor cells for eight days in static culture *in vitro* and reported good level of cells proliferation and differentiation.

However, unlike Kandner and Hoerstrup groups' findings, multiple hydrolytic breakages and fragmentation reported Dvorin's group didn't mention anything about degradation performance of the scaffold of the heart valve.

7.6. Summary of ECM Results

As it stated in Chapter 6, to generate the correct ECM, the scaffold must be conditioned to mimic the physiological conditions of *in vivo.* There are two possible techniques to achieve this, namely *Biochemical and Biomechanical stimulations* and those are briefly discussed and reintroduced below.

7.6.1. Biochemical Stimulations of ECM

The production of ECM using biochemical stimulation was investigated by Sutherland et al. (2005). The authors carried out study using bone marrow mesenchymal stem cells (MSCs) on a combination of PGA/PLA scaffold seeded with cells with 20 µg/ml basic fibroblast growth factor and cultured over a period of four weeks, which was later on placed in pulmonary position in juvenile sheep for four and eight months. This *in vivo* culture resulted in the formation of collagen, GAG and only trace of elastin (Sutherland et al., 2005). Schmidt et al. (2006) undertook a study that used endothelial progenitor cells from chorionic villi and were differentiated into endothelial cells using various hormones and growth factors including vascular endothelial growth factor, human fibroblasts growth factor, human recombinant long-insulin-like growth factor – I, human epidermal growth factor, and 2% fetal bovine serum on PGA dip-coated with P4HB. After 21 days of culture under both mechanical strain and additional seven days of mechanical conditioning, there was ECM formation and EC differentiation (Schmidt et al., 2006). Schmidt's group, in 2007, used fetal human amniotic progenitors (single-cell source) and seeded them onto PGA biomaterial dip-coated in P4HB and provided these cells with same cells and growth factors, hydrocortisone, heparin, ascorabte and 20% fetal bovine serum to result in the production of 80% GAGs, 5% collagen and no elastin (Schmidt et al., 2007). In 2006, Schmidt et al. tried to compare the effect of biochemical stimulation (by addition of hormones and other growth factors) and mechanical stimulation and found that biochemical stimulation resulted in the production of 16% collagen, 28% GAG and no elastin. Mechanical stimulation by itself, however, did not result in the production of any ECM components which was quite surprising at the time.

7.6.2. Mechanical Stimulation

Regardless of Schmidt et al.'s findings, cells within the cardiovascular tissues have been shown to respond to mechanical stimuli by controlling the synthesis of almost all important components of the ECM (Gupta and Grando-Allen, 2006). In addition, earlier and various others studies have shown that mechanical stimulation plays an important role not only in tissue engineering of heart valve but also in various applications like the osteochondral

Concept and Development of Tissue Engineering ... 153

defects, wherein it played an important role in MSC dispersal, proliferation, differentiation and cell death (Kelly and Prendergast, 2005). Moreover, the use of hormones or steroids to induce the production of ECM components can be dominant, and various studies have shown that cells are able to produce ECM components like collagen, GAG, elastin, matrix metalloproteinase (MMPs) by mechanical simulation. Additionally, the use of cardiac fibroblasts, cardiomyocytes and smooth muscle cells can result in increased the production of collagen, MMPs and collagen, elastin and MMPs, respectively, under cyclic strain (Gupta and Grando-Allen, 2006).

Table 7.5. Stimulations of ECM

Method used	Result obtained			Authors
	Collagen (% of native)	GAGs (% of native)	Elastin (% of native)	
Biochemical stimulation	Produced but % not given	Produced but % not given	Just detectable (*in vivo*)	Sutherland *et al.* (2005)
Biochemical stimulation	Produced but % not given	Produced but % not given		Schmidt *et al.* (2006)
Biochemical stimulation	5	80		Schmidt *et al.* (2007)
Biochemical stimulation	16	28		Schmidt *et al.* (2006)
Mechanical stimulation	85	60		Hoerstrup *et al.* (2000)
Mechanical stimulation	25	37		Hoerstrup *et al.* (2002)
Mechanical stimulation	Produced but % not given	Produced but % not given		Engelmayr *et al.* (2005)

Studies carried out by Hoerstrup et al. (2000), showed the formation of 85% collagen, 60% GAG and no elastin after 28 days of culture of myofibroblasts on PGA dip-coated with P4HB under mechanical stimulation *in vitro* (Hoerstrup et al., 2000). In 2002, Hoerstrup's group produced 25% collagen and 37% GAG by the use of human bone marrow MSC on similar biomaterial as used previously (Hoerstrup et al., 2002). Engelmayr (2005) used ovine smooth muscle cells on PGA/PLLA (50:50) and subjected to correct physiological cyclic conditions for three weeks to obtain the generation of collagen and GAG (Engelmayr et al., 2005). All these studies emphasised the increase in strength of the biomaterial/tissue formed due to mechanical stimulation, which is very important for the required application. Table 8 gives summary of the ECM results using different simulations techniques.

References

Agrawal, C. M., and R. B. Ray, Biodegradable polymeric scaffolds for musculoskeletal tissue engineering, *Journal of Biomedical Materials Research,* 2001. 55(2): pp. 141-150.

Ameer, G. A., T. A. Mahmood and R. Langer, A biodegradable composite scaffold for cell transplantation, *Journal of Orthopaedic Research,* 2002. 20(1): pp. 16-19.

Atala, A., Tissue engineering of human bladder, *Br. Med. Bull.,* 2011. 97(1): pp. 81-104.

Bader, A., T. Schiling, O. E. Teebken, G. Brandes, T. Herden, G. Steinhoff and A. Haverich, Tissue engineering of heart valves—Human endothelial cell seeding of detergent acellularized porcine valves, *Eur. J. Cardiothorac. Surg.,* 1998. 14(3): pp. 279-84.

Bertipaglia, B., F. Ortolani, L. Petrelli, G. Gerosa, M. Spina, P. Pauletto, D. Casarotto, M. Marchini and S. Sartore, Cell characterization of porcine aortic valve and decellularized leaflets repopulated with aortic valve interstitial cells: the VESALIO Project (Vitalitate Exornatum Succedaneum Aorticum Labore Ingenioso Obtenibitur), *Ann. Thorac Surg.,* 2003. 75(4): pp. 1274-82.

Bhattarai, N., D. Edmondson,O. Veiseh, F. A. Matsen and M. Zhang, Electrospun chitosan-based nanofibres and their cellular compatibility, *Biomaterials,* 2005. 26(31): pp. 6176-6184.

Booth, C., S. A. Korossis, H. E. Wilcox, K. G. Watterson, J. N. Kearney, J. Fisher and E. Ingram, Tissue engineering of cardiac valve prostheses I: development and histological characterization of an acellular porcine scaffold, *J. Heart Valve Dis.,* 2002. 11(4): pp. 457-62.

Brody, S. and A. Pandit, Approaches to heart valve tissue engineering scaffold design, *J. Biomed. Mater Res. B Appl. Biomater.,* 2007. 83(1): pp. 16-43.

Chen, R., Y. Morsi, S. Patel, Q. F. Ke and X. M. Mo, A novel approach via combination of electrospinning and FDM for tri-leaflet heart valve scaffold fabrication, *Frontiers of Materials Science in China,* 2009. 3(4), pp. 359-366.

Courtney, T., M. S. Sacks, J. Stankus, J. Guan and W. R. Wagner, Design and analysis of tissue engineering scaffolds that mimic soft tissue mechanical anisotropy, *Biomaterials,* 2006. 27(19): pp. 3631-3638.

Curtil, A., D. E. Pegg and A. Wilson, Repopulation of freeze-dried porcine valves with human fibroblasts and endothelial cells, *J. Heart Valve Dis.,* 1997. 6(3): pp. 296-306.

Cuy, J. L., B. L. Beckstead, C. D. Brown, A. S. Hoffman and C. M. Giachelli, Adhesive protein interactions with chitosan: Consequences for valve endothelial cell growth on tissue-engineering materials, *J. Biomed. Mater Res. A.,* 2003. 67(2): pp. 538-47.

De Cock, L.J., De Cock, L.J., De Koker, S., De Vos, F., Vervaet, C., Remon, J.P., and De Geest, B.G.,Layer-by-Layer Incorporation of Growth Factors in Decellularized Aortic Heart Valve Leaflets. *Biomacromolecules,* 2010. 11(4): p. 1002-1008.

Deng, C., et al., Application of decellularized scaffold combined with loaded nanoparticles for heart valve tissue engineering. *Journal of Huazhong University of Science and Technology—Medical Sciences—,* 2011. 31(1): p. 88-93.

Dohmen, P.M., et al., A tissue engineered heart valve implanted in a juvenile sheep model. Medical science monitor : *international medical journal of experimental and clinical research,* 2003. 9(4): p. BR97-BR104.

Dvorin, E.L., Wylie-Sears, J., Kaushal, S., Martin, D.P., and Bischoff, J., Quantitative Evaluation of Endothelial Progenitors and Cardiac Valve Endothelial Cells: Proliferation and Differentiation on Poly-glycolic acid/Poly-4-hydroxybutyrate Scaffold in Response to Vascular Endothelial Growth Factor and Transforming Growth Factor β1. *Tissue Engineering,* 2003. 9(3): p. 487-493.

Engelmayr Jr, G.C., Rabkin, E., Sutherland, F.W.H., Schoen, F.J., Mayer Jr, J.E., and Sacks, M.S., The independent role of cyclic flexure in the early in vitro development of an engineered heart valve tissue. *Biomaterials,* 2005. 26(2): p. 175-187

Fang, N.T., Xie, S.Z., Wang, S.M., Gao, H.Y., Wu, C.G., and Pan, L.F., Construction of tissue-engineered heart valves by using decellularized scaffolds and endothelial progenitor cells. *Chinese Medical Journal,* 2007. 120(8): p. 696-702.

Geffre CP, Margolis DS, Ruth JT, DeYoung DW, Tellis BC, Szivek JA. "A novel biomimetic polymer scaffold design enhances bone ingrowth" *J. Biomed. Mater Res. A.* 2009; 91(3):795-805.

Gilbert, T.W., Sellaro, T.L., and Badylak, S.F., Decellularization of tissues and organs. *Biomaterials,* 2006. 27(19): p. 3675-3683

Gonzalez, C. A. G., A. V. Gonzalez, A. M. L. Periago, P. S. Paternault and C. Domingo, Composite fibrous biomaterials for tissue engineering obtained using a supercritical CO2 antisolvent process, *Acta Biomater.,* 2009. 5(4): pp. 1094-103.

Grauss, R.W., Hazekamp, M.G., Oppenhuizen, F., Van Munsteren, C.J., Gittenberger-De Groot, A.C., and DeRuiter, M.C.,Histological evaluation of decellularised porcine aortic valves: matrix changes due to different decellularisation methods. *European Journal of Cardio-Thoracic Surgery,* 2005. 27(4): p. 566-571.

Hoerstrup, S. P., Zund, G., Schoeberlein, A., Ye, Q., Vogt, P. R., and Turina, M. I.. Fluorescence activated cell sorting: a reliable method in tissue engineeing of a bioprosthetic valve. *Annals of thoracicsurgery,* 1998.66, 1653–1657.

Hoerstrup, S. P., A. Kander, S. Melnitchouk, A. Trojan, K. Eid, J. Tracy, R. Sodian, J. F. Visjager, S.A. Kolb, J. Grunenfelder, G. Zund and M. I. Turina, Tissue Engineering of Functional Trileaflet Heart Valves From Human Marrow Stromal Cells, *Circulation,* 2002. 106(12 Suppl 1): pp. 1143-50.

Hoerstrup, S. P., R. Sodian, S. Daebritz, J. Wang, E. A. Bacha, D. P. Martin, A. M. Moran, K. J. Guleserian, J. S. Sperling, S. Kaushal, J. P. Vacanti, F. J. Schoen, J. E. Mayer, Jr., Functional living trileaflet heart valves grown in vitro, *Circulation,* 2000. 102(19 Suppl 3): pp. III44-9.

Hollister, S. J., Porous scaffold design for tissue engineering, *Nature Materials,* 2005. 4(7): pp. 518-524.

Hong, H., Dong, N., Shi, J., Chen, S., Guo, C., Hu, P., and Qi, H.,Fabrication of a Novel Hybrid Heart Valve Leaflet for Tissue Engineering: An In vitro Study. *Artificial Organs,* 2009. 33(7): p. 554-558.

Hutmacher, D. W., Scaffold design and fabrication technologies for engineering tissues—state of the art and future perspectives, *J. Biomater Sci. Polym.* Ed., 2001. 12(1): pp. 107-24.

Iop, L., Renier, V., Naso, F., Piccoli, M., Bonetti, A., Gandaglia, A., Pozzobon, M., Paolin, A., Ortolani, F., Marchini, M., Spina, M., De Coppi, P., Sartore, S., and Gerosa, G., The influence of heart valve leaflet matrix characteristics on the interaction between human mesenchymal stem cells and decellularized scaffolds. *Biomaterials,* 2009. 30(25): p. 4104-4116.

Jockenhoevel, S., K. Chalabi, J. S. Sachweh, H. V. Groesdonk, L. Demircan, M. Grossmann, G. Zund and B. J. Messmer, Tissue engineering: Complete autologous valve conduit—*A new moulding technique, Thoracic and Cardiovascular Surgeon,* 2001a. 49(5): pp. 287-290.

Jockenhoevel, S., G. Zund, S. P. Hoerstrup, K. Chalabi, J. S. Sachweh, L. Demircan, B. J. Messmer and M. Turina, Fibrin gel-advantages of a new scaffold in cardiovascular tissue engineering, *Eur. J. Cardiothorac. Surg.,* 2001b. 19(4): pp. 424-30.

Kadner, A., Hoerstrup, S.P., Tracy, J., Breymann, C., Maurus, C.h.F., Melnitchouk, S., Kadner, G., Zund, G., and Turina, M., Human umbilical cord cells: a new cell source for cardiovascular tissue engineering. *The Annals of Thoracic Surgery,* 2002. 74(4): p. 1422-1428.

Kaushal, S., Amiel, G.E., Guleserian, K.J., Shapira, O.M., Perry, T., Sutherland, F.W., Rabkin, E., Moran, A.M., Schoen, F.J., Atala, A., Soker, S., Bischoff, J., and Mayer, J.E., Jr., Functional small-diameter neovessels created using endothelial progenitor cells expanded ex vivo. *Nature Medicine,* 2001. 7(9): p. 1035-40.

Karageorgiou, V., and D. Kaplan, Porosity of 3D biomaterial scaffolds and osteogenesis, *Biomaterials,* 2005. 26(27): pp. 5474-5491.

Kasimir, M.T., Weigel, G., Sharma, J., Rieder, E., Seebacher, G., Wolner, E., and Simon, P., The decellularized porcine heart valve matrix in tissue engineering: platelet adhesion and activation. *Thrombosis and Haemostasis,* 2005. 94(3): p. 562-7.

Kim, W. G., S. K. Cho, M. C. Kang, T. Y. Lee and J. K. Park, Tissue- engineered heart valve leaflets: An animal study, *International Journal of Artificial Organs,* 2001. 24(9): pp. 642-648.

Kim, W. G., J. K. Park and W. Y. Lee, Tissue-engineered heart valve leaflets: an effective method of obtaining acellularized valve xenografts, *Int. J. Artif. Organs.,* 2002. 25(8): pp. 791-7.

Lee, K. Y. and D. J. Mooney, Hydrogels for tissue engineering, *Chem. Rev.,* 2001. 101(7): pp. 1869-79.

Lee, K., E. A. Silva and D. J. Mooney, Growth factor delivery-based tissue engineering: general approaches and a review of recent developments, *J. R. Soc. Interface.,* 2011. 8(55): pp. 153-70.

Leyh, R. G., M. Wilhelmi, T. Walles, K. Kallenbach, P. Rebe, A. Oberbeck, T. Herden, A. Haverich and H. Mertsching, Acellularized porcine heart valve scaffolds for heart valve tissue engineering and the risk of cross-species transmission of porcine endogenous retrovirus, *J. Thorac. Cardiovasc. Surg.,* 2003. 126(4): pp. 1000-4.

Liao, J., E.M. Joyce, and M.S. Sacks, Effects of decellularization on the mechanical and structural properties of the porcine aortic valve leaflet. *Biomaterials,* 2008. 29(8): p. 1065-1074.

Liu, X., and P. X. Ma, Polymeric scaffolds for bone tissue engineering, *Annals of Biomedical Engineering,* 2004. 32(3): pp. 477-486.

Lieshout, M. I. V., C. M. Vaz, M. C. Rutten, G. W. Peters and F. P. Baaijens, Electrospinning versus knitting: two scaffolds for tissue engineering of the aortic valve, *J. Biomater. Sci. Polym. Ed.,* 2006. 17(1-2): pp. 77-89.

Lichtenberg, A., et al., In vitro re-endothelialization of detergent decellularized heart valves under simulated physiological dynamic conditions. *Biomaterials,* 2006. 27(23): p. 4221-4229.

Lü, X., et al., Cross-linking effect of Nordihydroguaiaretic acid (NDGA) on decellularized heart valve scaffold for tissue engineering. *Journal of Materials Science: Materials in Medicine,* 2010. 21(2): p. 473-480.

Maher, P. S., R. P. Keatch, K. Donnelly, R. E. MacKay and J. Z. Paxton, Construction of 3D biological matrices using rapid prototyping technology, *Rapid Prototyping Journal,* 2009. 15(3): pp. 204-210.

Masters, K. S., D. N. Shah, G. Walker, L. A. Leinwand and K. S. Anseth, Designing scaffolds for valvular interstitial cells: Cell adhesion and function on naturally derived materials, *Journal of Biomedical Materials Research Part A,* 2004. 71A(1): pp. 172-180.

Mei, N., G. Chen, P. Zhou, X. Chen, Z. Z. Shao, L. F. Pan and C. G. Wu, Biocompatibility of poly(epsilon-caprolactone) scaffold modified by chitosan—The fibroblasts proliferation in vitro, *Journals of Biomaterials Applications,* 2005. 19(4): pp. 323-339.

Mendelson, K., and F. J. Schoen, Heart valve tissue engineering: concepts approaches, progress, and challenges, *Ann. Biomed. Eng.,* 2006. 34(12): pp. 1799-819.

Mol, A., Functional tissue engineering of human heart valve leaflets. *Phd Thesis,* 2005: Technische Universiteit Eindhoven

Morsi, Y. S., and I. Birchall, Tissue engineering a functional aortic heart valve: an appraisal, *Future Cardiology,* 2005. 1(3): pp. 405-411.

Morsi, Y. S., I. E. Birchall and F. L. Rosenfeldt., Artificial aortic valves: an overview, *Int. J. Artif. Organs,* 2004. 27(6): pp. 445-51.

Neidert, M. R., and R. T. Tranquillo, Tissue-engineered valves with commissural alignment, *Tissue Engineering,* 2006. 12(4): pp. 891-903.

Nuttelman, C. R., S. M. Henry and K. S. Anseth, Synthesis and characterization of photocross-linkable, degradable poly(vinyl alcohol)-based tissue engineering scaffolds, *Biomaterials,* 2002. 23(17): pp. 3617-3626.

Novakovic, G. V. and I. R. Freshney, Culture of cells for tissue engineering, eBook, Chapter 6, J. Wiley, 2006. [Viewed from: http://onlinelibrary.wiley.com/doi/10.1002 /04717 41817.ch6/summary] on 26/06/11.

O'Brien, M. F., S. Goldstein, S. Walsh, K. S. Black, R. Elkins and D. Clarke, The SynerGraft valve: a new acellular (nonglutaraldehyde-fixed) tissue heart valve for autologous recellularization first experimental studies before clinical implantation, *Semin. Thorac. Cardiovasc. Surg.,* 1999. 11(4 Suppl 1): pp. 194-200.

Oliveira, J. M., M. T. Rodrigues, S. S. Silva, P. B. Malafaya, M. E. Gomes, C. A. Viegas, I. R. Dias, J. T. Azevedo, J. F. Mano and R. L. Reis, Novel hydroxyapatite/chitosan bilayered scaffold for osteochondral tissue-engineering applications: Scaffold design and its performance when seeded with goat bone marrow stromal cells, *Biomaterials,* 2006. 27(36): pp. 6123-6137.

Perry, T.E., Kaushal, S., Sutherland, F.W.H., Guleserian, K.J., Bischoff, J., Sacks, M., and Mayer, J.E., Bone marrow as a cell source for tissue engineering heart valves. *The Annals of Thoracic Surgery,* 2003. 75(3): p. 761-767.

Pham, Q. P., U. Sharma, and A. G. Mikos, Electrospinning of polymeric nanofibres for tissue engineering applications: A review, *Tissue Engineering,* 2006. 12(5): pp. 1197-1211.

Prabha S, and S. Verghese, Existence of proviral porcine endogenousretrovirus in freshand decellularised porcine tissues, *Ind. J. Med. Microbiol.* 2008. 26(3),pp. 228-232.

Rabkin, E., S. P. Hoerstrup, M. Aikawa, J. E. Mayer Jr, and F. J. Schoen, Evolution of cell phenotype and extracellular matrix in tissue-engineered heart valves during in-vitro maturation and in-vivo remodeling, *Journal of Heart Valve Disease,* 2002. 11(3): pp. 308-314.

Ramamurthi, A., and I. Vesely, Evaluation of the matrix-synthesis potential of cross-linked hyaluronan gels for tissue engineering of aortic heart valves, *Biomaterials,* 2005. 26(9): pp. 999-1010.

Rieder, E., M. T. Kasimir, G. Silberhumer, G. Seebacher, E. Wolner, P. Simon and G. Weigel, Decellularization protocols of porcine heart valves differ importantly in efficiency of cell removal and susceptibility of the matrix to recellularization with human vascular cells, *The Journal of Thoracic and Cardiovascular Surgery*, 2004. 127(2): pp. 399-405.

Rothenburger, M., P. Vischer, W. Volker, B. Glasmacher, E. Berendes, H. H. Scheld and M. Deiwick, In vitro modelling of tissue using isolated vascular cells on a synthetic collagen matrix as a substitute for heart valves, *Thorac Cardiovasc. Surg.*, 2001. 49(4): pp. 204-9.

Rothenburger, M., W. Volker, P. Vischer, B. Glasmacher, H. H. Scheld and M. Deiwick, Ultrastructure of proteoglycans in tissue-engineered cardiovascular structures, *Tissue Eng.*, 2002. 8(6): pp. 1046-56.

Samouillan, V., J. Dandurand-Lods, A. Lamure, E. Maurel, C. Lacabanne, G. Gerosa, A. Venturini, D. Casarotto, L. Gherardini and M. Spina, Thermal analysis characterization of aortic tissues for cardiac valve bioprostheses, *J. Biomed. Mater Res.*, 1999. 46(8): pp. 531-8.

Schaefermeier, P. K., D. Szymanski, F. Weiss, P. Fu, T. Lueth, C. Schmitz, B. M. Meiser, B. Reichart and R. Sodian, Design and Fabrication of Three-Dimensional Scaffolds for Tissue Engineering of Human Heart Valves, *European Surgical Research*, 2009. 42(1): pp. 49-53.

Schenke-Layland, K., Riemann, I., Opitz, F., König, K., Halbhuber, K.J., and Stock, U.A.,Comparative study of cellular and extracellular matrix composition of native and tissue engineered heart valves. *Matrix Biology*, 2004. 23(2): p. 113-125.

Schense, J. C., and J. A. Hubbell, Cross-Linking Exogenous Bifunctional Peptides into Fibrin Gels with Factor XIIIa, *Bioconjugate Chemistry*, 1998. 10(1): pp. 75-81.

Schleicher, M., Wendel, H.P., Fritze, O., and Stock, U.A., In vivo tissue engineering of heart valves: evolution of a novel concept. *Regenerative Medicine*, 2009. 4(4): p. 613-619.

Schmidt, D., Breymann, C., Weber, A., Guenter, C.I., Neuenschwander, S., Zund, G., Turina, M., and Hoerstrup, S.P., Umbilical Cord Blood Derived Endothelial Progenitor Cells for Tissue Engineering of Vascular Grafts. *Annals of Thoracic Surgery*, 2004. 78(6): p. 2094-2098.

Schmidt, D. and Hoerstrup, S.P., Tissue-engineered heart valves based on human cells. *Swiss Medical Weekly*, 2006. 136(39-40): p. 618-623.

Schnell, A. M., Hoerstrup, S. P., Zund, G., Kolb, S., Sodian, R., Visjager, J. F., Grunenfelder, J., Suter, A., andTurina, M.. Optimal cell source for tissue engineering: Venous vs. aortic human myofibroblasts. *Thoracic and cardiovascular surgeon*, 2001, 49(4), 221–225.

Seebacher, G., Grasl, C., Stoiber, M., Rieder, E., Kasimir, M.T., Dunkler, D., Simon, P., Weigel, G., and Schima, H.., Biomechanical Properties of Decellularized Porcine Pulmonary Valve Conduits. *Artificial Organs*, 2008. 32(1): p. 28-35.

Shah, D. N., S. M. R. Work and K. S. Anseth, The effect of bioactive hydrogels on the secretion of extracellular matrix molecules by valvular interstitial cells, *Biomaterials*, 2008. 29(13): pp. 2060-72.

Shi, Y., L. Rittman and I. Vesely, Novel geometries for tissue-engineered tendonous collagen constructs, *Tissue Engineering*, 2006. 12(9): pp. 2601-2609.

Shinoka, T., Tissue-engineered heart valves: Autologous cell seeding on biodegradable polymer scaffold, *Artificial Organs*, 2002. 26(5): pp. 402-406.

Shinoka, T., C. K. Breuer, R. E. Tanel, G. Zund, T. Miura, P. X. Ma, R. Langer, J. P. Vacanti and J. E. Mayer, Jr., Tissue engineering heart valves: Valve leaflet replacement study in a lamb model, *Annals of Thoracic Surgery,* 1995. 60(Suppl 3): S513-S516.

Shinoka, T., P. X. Ma, D. S. Tim, C. K. Breuer, R. A. Cusick, G. Zund, R. Langer, J. P. Vacanti and J. E. Mayer, Jr., Tissue-engineered heart valves: Autologous valve leaflet replacement study in a lamb model, *Circulation,* 1996. 94(9 Suppl): pp. II164-II168.

Simon, P., M. T. Kasimir, G. Seebacher, G. Weigel, R. Ullrich, U. S. Muhar, E. Rieder and E. Wolner, Early failure of the tissue engineered porcine heart valve SYNERGRAFT in pediatric patients, *Eur. J. Cardiothorac Surg.,* 2003. 23(6): pp. 1002-6.

Sodian R., S. P. Hoerstrup, J. S. Sperling, S. Daebritz, D. P. Martin, A. M. Moran, B. S. Kim., F. J. Schoen, J. P. Vacanti and J. E. Mayer, Jr., Early in vivo experience with tissue-engineered trileaflet heart valves, *Circulation,* 2000a. 102(19 Suppl 3): pp. III22-9.

Sodian, R., S. P. Hoerstrup, J. S. Sperling, S. H. Daebritz, D. P. Martin, F. J. Schoen, J. P. Vacanti and J. E. Mayer, Tissue engineering of heart valves: In vitro experiences, *Annals of Thoracic Surgey,* 2000b. 70(1): pp. 140-144.

Sodian, R., S. P. Hoerstrup, J. S. Sperling, D. P. Martin, S. Daebritz, J. E. Mayer Jr. and J. P. Vacanti, Evaluation of biodegradable, three-dimensional matrices for tissue engineering of heart valves, *ASIAO Journal,* 2000c. 46(1): pp. 107-110.

Sodian, R., J. S. Sperling, D. P. Martin, A. Egozy, U. Stock, J. E. Mayer, Jr. and J. P. Vacanti, Fabrication of a trileaflet heart valve scaffold from a polyhydroxyalkanoate biopolyester for use in tissue engineering, *Tissue Eng.,* 2000d. 6(2): pp. 183-8.

Sodian, R., M. Loebe, A. Hein, D. P. Martin,S. P. Hoerstrup, E. V. Potapov, H. Hausmann, T. Lueth and R. Hetzer, Application of stereolithography for scaffold fabrication for tissue engineered heart valves, *ASIAO Journal,* 2002. 48(1): pp. 12-16.

Spina, M., Ortolani, F., El Messlemani, A., Gandaglia, A., Bujan, J., Garcia-Honduvilla, N., Vesely, I., Gerosa, G., Casarotto, D., Petrelli, L., and Marchini, M.,Isolation of intact aortic valve scaffolds for heart-valve bioprostheses: Extracellular matrix structure, prevention from calcification, and cell repopulation features. *Journal of Biomedical Materials Research Part A,* 2003. 67A(4): p. 1338-1350.

Stamm, C., Khosravi, A., Grabow, N., Schmohl, K., Treckmann, N., Drechsel, A., Nan, M., Schmitz, K.P., Haubold, A., and Steinhoff, G.,Biomatrix/Polymer Composite Material for Heart Valve Tissue Engineering. *Ann. Thorac. Surg,* 2004. 78(6): p. 2084-2093.

Steinhoff, G., U. Stock, N. Karim, H. Mertsching, A.Timke, R. R. Meliss, K. Pethig, A. Haverich and A. Bader, Tissue engineering of pulmonary heart valves on allogenic acellular matrix conduits: In vivo restoration of valve tissue, *Circulation,* 2000. 102(9): pp. III50-III55.

Suh, J. K. F., and H. W. T. Matthew, Application of chitosan-based polysaccharide biomaterials in cartilage tissue engineering: a review, *Biomaterials,* 2000. 21(24): pp. 2589-2598.

Sutherland, F.W.H., Perry, T.E., Yu, Y., Sherwood, M.C., Rabkin, E., Masuda, Y., Garcia, G.A., McLellan, D.L., Engelmayr, G.C., Sacks, M.S., Schoen, F.J., and Mayer, J.E., From Stem Cells to Viable Autologous Semilunar Heart Valve. *Circulation,* 2005. 111(21): p. 2783-2791.

Taylor, P. M., S. P. Allen, S. A. Dreger and M. H.Yacoub, Human cardiac valve interstitial cells in collagen sponge: a biological three-dimensional matrix for tissue engineering, *J. Heart Valve Dis.,* 2002. 11(3): pp. 298-306.

Tudorache I,Cebotari, S., Sturz, G., Kirsch, L., Hurschler, C., Hilfiker, A., Haverich, A., and Lichtenberg, A., Tissue engineering of heart valves: biomechanical and morphological properties of decellularized heart valves. *J. Heart Valve Dis,* 2007. 16(5): p. 567-573.

Vesely, I. The role of elastin in aortic valve mechanics. *Journal of Biomechanics,* 1998, 31(2): 115-123.

Wu K.., Y. L. Liu, B. Cui and Z. Han, Application of stem cells for cardiovascular grafts tissue engineering, *Transplant Immunology,* 2006. 16(1): pp. 1-7.

Yang, S., K. F. Leong, Z. Du and C. K.Chua, The design of scaffolds for use in tissue engineering. Part I. Traditional factors, *Tissue Eng.,* 2001. 7(6): pp. 779-89.

Yang, T. L., Chitin-based materials in tissue engineering: Applications in soft tissue and epithelial organ, *International Journal of Molecular Sciences,* 2011. 12(3): pp. 1936-1963.

Yamanami, M., Yahata, Y., Uechi, M., Fujiwara, M., Ishibashi-Ueda, H., Kanda, K., Watanabe, T., Tajikawa, T., Ohba, K., Yaku, H., and Nakayama, Y., Development of a Completely Autologous Valved Conduit With the Sinus of Valsalva Using In-Body Tissue Architecture Technology. *Circulation,* 2010. 122(11 suppl 1): p. S100-S106

Yarlagadda, P.K.D.V., Chandrasekharan, M., and Shyan, J.Y.M., Recent advances and current developments in tissue scaffolding. *Biomedical Materials and Engineering,* 2005. 15(3): p. 159-177.

Ye, Q., G. Zund, P. Benedikt, S. Jockenhoevel, S. P. Hoerstrup, S. Sakyama, J. A. Hubbell and M. Turina, Fibrin gel as a three dimensional matrix in cardiovascular tissue engineering, *Eur. J. Cardiothorac. Surg,* 2000. 17(5): pp. 587-591.

Zeltinger, J., L. K. Landeen, H. G. Alexander, I. D. Kidd and B. Sibanda, Development and characterization of tissue-engineered aortic valves, *Tissue Eng.,* 2001. 7(1): pp. 9-22.

Zhou, J., Fritze, O., Schleicher, M., Wendel, H.P., Schenke-Layland, K., Harasztosi, C., Hu, S., and Stock, U.A., Impact of heart valve decellularization on 3D ultrastructure, immunogenicity and thrombogenicity. *Biomaterials,* 2010. 31(9): p. 2549-2554.

Zong, X., H. Bien, H., C. Y. Chung, L. Yin, D. Fang, B. S. Hsiao, B. Chu and E. Entcheva, Electrospun fine-textured scaffolds for heart tissue constructs, *Biomaterials,* 2005. 26(26): pp. 5330-8.

Zund, G., C. K. Breuer, T. Shinoka, P. X. Ma, R. Langer, J. E. Mayer and J. P. Vacanti, The in vitro construction of a tissue engineered bioprosthetic heartvalve, *Eur. J. Cardiothorac. Surg.* 1997. 11(3): pp. 493-7.

Zund, G., Hoerstrup, S. P., Schoeberlein, A., Lachat, M., Uhlschmid, G., Vogt, P. R., and Turina, M. Tissue engineering: a new approach in cardiovascular surgery; seeding of human fibroblasts followed by human endothelial cells on a resorbable mesh. *European journal of cardiothoracic surgery,* 1998. 13, 160–164.

Chapter VIII

Closing Remarks
and Future Challenges

8.1. Current Issues

Though the first mechanical and bioprosthetic valve replacements were introduced over 40 years ago, tissue engineering of heart valve is still in the early development (Chandran et al., 2006). Subsequently, the advances in tissue engineering (TE) of heart valves have been moderate over the past few years with respect to creating one that is clinically applicable. Nevertheless, an abundance of knowledge has been gained in various areas pertinent to TE of heart valves. These include extracellular matrix production (Dreger et al., 2006; Stamm et al., 2004), cellular interactions, including stem cells (Alperin et al., 2005; Liechty et al., 2000) and haemodynamic characterization, for example, the effect of shear stress on cells (Butcher et al., 2004; Feugier et al., 2005). Nevertheless, in general, data obtained so far need verifications and validation in vivo before the clinical implementation of TEHV can take place. Although, substantial achievements have been accomplished to date, much exertion still needed, both in defining the functional requirements of a TEHV suitable for clinical evaluation, as well as in developing the scientific techniques to create functional one.

This book introduces both current valve replacements, namely, artificial heart valves, and the futuristic tissue engineering heart valves. Both types of valves have gone through a number of developments, and the effort to optimize these valves is still ongoing. These investigations have highlighted a number of positive outcomes but also a number of future challenges. This chapter briefly discusses the current research programs and the future directions to create a viable valve substitute to alleviate the problem of HVD.

As stated in Chapter 3, the existing heart valve substitutes are of two principal types: mechanical prosthetic valves with components manufactured of non-biologic materials (e.g., carbon, metal, polymeric materials) and tissue valves that are formulated of either from animals or human tissues, in part. The main drawbacks of mechanical valves are related to the considered risk of systemic thromboembolism and thrombosis (largely due to abnormal flow past rigid occluders and the use of non-physiologic surfaces), and serious complications from hemorrhage associated with the necessary use of chronic anti-coagulation therapy (Schoen and Levy, 1999; Meuris, 2007). Although these valves offer superior long-term durability, essentially lasting the lifetime of the recipient, mechanical valves require chronic anticoagulation therapy to reduce the potential for thromboembolism and are susceptible to infection. These limitations are of particular concern in women of childbearing age, who carry an elevated risk of valve-related thrombosis and anticoagulant-related embryopathy during pregnancy (Van Mook and Peeters, 2005).

Artificial bioprosthetic tissue valves, although they maintain a low rate of thromboembolism without anti-coagulation therapy suffer from calcification that contributes to the deterioration of cuspal tissue, leading to regurgitation through tears in calcified leaflets. Moreover, non-calcified destruction to the valvular structure that could accrue through constrained abnormal valve motion is also a major mechanism of degradation in porcine and pericardial prosthetic valves. Moreover, this calcific and fatigue-related damage occurs over a relatively short time frame of 10 to 15 years, resulting either in surgical intervention or mortality. This highlighted the critical need to sustain the integrity of the tissue structure so that the lifespan of the valve can be extended (Engelmayr, 2005).

However, irrespective of the mechanisms that underlie the problem of the damaged leaflets in bioprosthetic valves, there is still the concern of their inability to simulate the physical properties of natural leaflets. Now, it is recognised that fixation of tissues via glutaraldehyde causes soluble proteins to be incorporated into the structural proteins lead to an increase of structural instability and rigidity. Flexibility of the leaflets is crucial for maintaining the correct functionality of the leaflets. In general, however, tissue valves are normally used for elderly patients, and as the aging population increases, the emphasis of improving tissue valves increases. Homograft valves from human cadavers provide best practice in relation to hemodynamics and functionality; however, their use is restricted by donor scarcity. Nevertheless, current investigation and attempts to enhance the biocompatibility and durability of artificial valves will likely bring some degree of success in the future (Engelmayr, 2005).

However, regardless of the degree of advancement in these types of valves, these artificial valves lack living cells and thus do not retain the capacity for growth that is conceptually appealing in reconstructing pediatric valvular lesions. Moreover, artificial replacements in general are based on a number of engineering and biological compromises, they are limited in their ability to fully integrate into the host, and they do not remodel with growing children. Even the use of autologous tissue or allografts is limited by donor site and accessibility, and there is always the question that using lifelong immuno-suppression is essential (Terada et al., 2000).

Tissue engineering (TE) of heart valves, on the other hand, is an innovative concept that has the potential to generate fully functional heart valves for patients with dysfunctional heart valves. In general design and development of any type of valve substitute, whether artificial or tissue engineered, one has to go through rigorous *in vitro* conditioning under physiological conditions followed by long-term animal trials. Researchers in the field normally adapt a systematic approach to develop the knowledge base, starting from bioengineering *in vitro* modelling of a conceptual design of a given biomaterial and structure, to cell culture and animals trials *in vivo*. Nevertheless, whether the concern is the development of artificial substitute or tissue engineering valve, the research and development in this field cover certain elements that are interrelated and that need to be explored, as shown in Figure 8.1 and briefly discussed below.

Naturally, depending on the final objective, some elements of the research shown above may be excluded. For example, if one is mainly concerned with decellularization approach, the materials development and manufacturing of the scaffold will not be an issue, but sterilization will be a main focus and so on. For polymer approach, the selection of biomaterials and manufacturing techniques of the scaffold as well as hemodynamic analysis are equally relevant and interrelated. However, if one is only concerned with the design of artificial heart valve, cell culture may not be an issue.

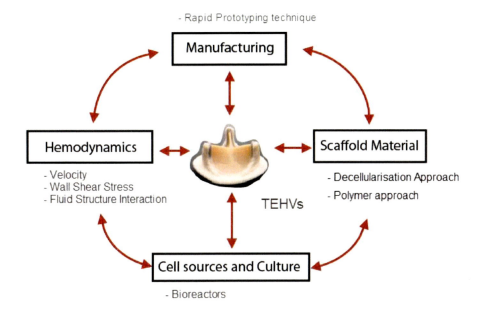

Figure 8.1. The interrelation of the research elements involved in design and optimization of heart valve.

As stated in Chapter 7, there are two approaches available in creating TEHV, namely decellularization and polymer approaches. However, Layland et al. (2003) pronounced that creating tissue engineering heart valve using decellularization method could prompt immunological concerns associated with inflammation and rejection, thus suggesting a new direction of an autologous method. The other polymer approach that recently enjoys tremendous approval is based on the design and optimization of polymer scaffold alone or in a combination with natural materials. This approach relies mainly on the design and development of the correct scaffold configuration that is made from an appropriate material that is capable of mimicking the nature valve and assists in the regeneration of the autologous ECM.

In a nutshell, it is recognized that both approaches for tissue engineering of heart valves have achieved considerable advancements in the last few decades. It is nevertheless indeterminate which one of these two techniques will be the optimum option in the future. Therefore, supplementary research for both approaches are still needed, and the following sections discuss the main current and future research issues that are required to build a knowledge base for tissue engineering heart valve research.

8.2. Bioengineering Developments

With the polymer approach, it is well recognized that the ideal scaffold structure for tissue engineering applications should be highly porous microstructure with the correct mechanical integrity and appropriate surface texture to enhance cell adhesion and proliferations. Moreover, it is recognised that for a given porosity, different microstructures could lead to different effective stiffness and permeability. Engineering techniques in the design and the optimization of the scaffold should include well-defined pores structure to maintain the required mechanical functionality and integrity, convection/diffusion mass transport properties, and manufacture of these structures within the complex 3D valve configurations (Morsi et al., 2011). For polymer heart valve scaffold, these issues are more multifaceted and mandate a more detailed analysis at the CAD modelling and before manufacturing and *in vitro* phases (Hollister, 2005).

As stated previously, the aortic valve consists of three leaflets that open and close during heart contraction and expansion. The maximum mechanical stresses experienced by the valve occur at points of maximum flexion, which include the point of leaflet attachment and the line of coaptation. During the cyclic functioning, the valve components experience stress variations as well as changes in the direction of curvature. The total stress on the valve becomes the sum of normal, shear and bending stresses (Thubrikar et al., 1982). Therefore, within the scaffold of the heart valve, the microstructure and the porosity to attain the desired mechanical and the physical properties of the conduit, including the sinuous, are different from that of the leaflets. Consequently, manufacturing the leaflets will require an optimisation process of the materials, thickness and surface structure as well as the degree of deformation and the coaptation that are different from the main body of the valve (Ramaswamy et al., 2010).

8.2.1. Experimental and Numerical Validations

The specific requirements from the polymer scaffold, to stand the correct physiological hemodynamic conditions during *in vitro* and *in vivo*, and the requirement that its structure and degree of deformation be of similar characteristics to the natural valve, demand a specific modelling approach. This approach requires a complete analysis and verifications of all the fluid forces and the resultant stress and strain in an anatomically correct aorta using various types of materials and operating conditions. As illustrated in Figure 8.2, to achieve this goal, different experimental and numerical techniques can be adapted individually or in combination, and these are discussed in the following sections.

8.2.1.1. Geometrical Design of the Valve

The dimensions and geometric relationships of the human aorta and the conceptual geometry of the valve are well defined in the literature. With respect to valve leaflets, there are a number of issues related to the design of them in terms of curvature, stiffness, texture, methods of attachments, etc. Initially, Mercer et al. developed the geometry of the human aortic leaflet via a moulding technique at 100 mm Hg of pressure (Mercer et al., 1973). Later, Swanson and Clark extracted the exact configuration of the whole valve by freezing the pig hearts under pressure for direct observation and measurements and reported the dimensional variability of normally functioning human aortic valve as a function of intra-aortic pressure between 0 and 120 mm Hg, respectively (Swanson and Clark, 1974). Recently, Kouhi (2012) optimised the work of Labrosse et al. (2006) on the design of valve leaflets and carried out a finite element analysis of a conceptual design of aortic valve in order to determine the analytical equations of three-dimensional geometric model of a tri-leaflet valve in both the open and closed configurations (Kouhi, 2012).

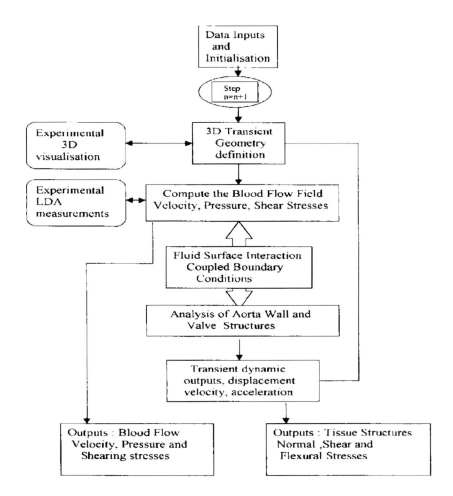

Figure 8.2. A typical hemodynamic research plan.

In general, however, by expending information available in literature, the initial geometry of the valve and structure could be modelled and visualised in 3D with the aid of CT using the CAD facilities, for example, Pro/Engineer CAD system. The Pro/Engineer is a powerful feature-based parametric solid modelling system with Pro/Surface and Pro/Mesh modules for design and analysis of parts with complex shapes. The solid models could then be transferred for meshing or grid generation using specialist commercial software required for optimisation of shape and performance using the numerical analysis of the fluid flow and structural analysis. Moreover, the geometry of the valve including the sinuses of valsalva and the correct structure of the leaflets need to be implemented to ensure the correct simulation of the physiological behaviour of the natural valve and to improve the operational of scaffold during *in vitro* and *in vivo* trials (Zund et al., 1997; Stock et al., 2000; Morsi and Birchall, 2006; Kouhi, 2012). These conditions are critical; in fact as early as 2000, Horestrup contributed regurgitation in polymer-based TEHVs to the enlargement of the valve conduit over time, which, in turn, causes inadequate closure of leaflets (Hoerstrup et al., 2000(a)). All the above-mentioned issues require a thorough investigation aiming at determine the optimum design of the valve and its leaflets using modelling and bioengineering approaches, as further discussed in the following section.

8.2.1.2. Hemodynamic and Structure Analysis

The mathematical description of the flow normally uses the unsteady, three-dimensional Navier-Stokes equations for an incompressible, generalized non-Newtonian fluid, with both laminar and turbulent flows normally considered. Turbulence can significantly affect the magnitude of the shear stresses and has been observed *in vitro* (Morsi et al., 2007b; Kouhi, 2012). Solutions to the Navier-Stokes equations have been published widely in this and other areas of research, such as the simulation of rheological processes. Both finite volume and finite element formulations have been used successfully (Norton and Sun, 2006; Kouhi, 2012). Major requirements here are the fine discretisation close to the curved moving boundaries and the accurate tight coupling of the surface forces and motions between the fluid and the structures. Hence, it is essential to consider both fluid and structural dynamics of the valve simultaneously in order for the valvular motion and the blood flow to be simulated correctly. This led to the inclusion of fluid-structure interaction (FSI) in the modelling of heart valve flow which was initially introduced by Peskin et al. and Morsi et al. (2007b), whereby a three-dimensional (3D) coupled fluid-structure dynamic model for a generic pericardial aortic valve in a rigid aortic root graft was developed and further highlighted the importance of FSI on the performance of heart valves. Blood flow was modelled as pulsatile, laminar, Newtonian, incompressible flow, and the structure model accounted for material and geometric non-linearties (Morsi et al., 2007b). Such analysis was later extened by Kouhi to include a complete analysis of the whole aortic valve and its leaflets under various cardiac conditions (Kouhi, 2012).

It should be noted that the structure of the tri-leaflet valve is the motion of the blood flow as the motion of thin-walled, leaflet-like structure drives quite complex. The large deformation of the physical (structural) boundary due to the fluid forces and non-linearity of the material increase the numerical difficulty considerably. Generally, the geometrically non-linear structure is customarily formulated in a Lagrangian description, while the fluid is often communicated in an Eulerian format due to the presence of convective term. Note that the presence of convective term in the case of changing boundary flow problem increases the numerical complexity sophistication in the Eulerian description, whereas the Lagrangian description for the structural relocating boundary is comparatively simpler. Moreover, due to the large flow alterations and large structural deformations, element disorder leads to the use of the Arbitrary Lagrangian- Eulerian (ALE) technique. The work presented in Morsi et al. (2007(b)) justifies the use of ALE technique for the type of those simulations in hand that involve mesh generation within an extensive time span. However, in the case of large deflection problem circumstances, the analysis requires some interpolation techniques to update the variables for newly generated mesh. The updated mesh may initiate an artificial diffusivity, which is a major upset regarding the applicability of ALE in some environments. Therefore, a fictitious domain/mortar element (FD/ME) approach has been recommended (Leveque and Li, 1994), whereby the fluid is described using a fixed mesh in an Eulerian positioning and a Lagrangian formulation for the solid. In the fictitious domain (FD) approach, the velocity constraint associated with the rigid internal boundaries is forced by modes of the Lagrange multiplier technique, while the mortar element (ME) method allows coupling of domains with dissimilar element distributions. The major advantage of the FD/ME method is that it does not necessitate any updating of the mesh of the fluid domain. However, updating of the mesh turns out to be inevitable for very large deflection circumstances (Morsi et al., 2007(b). All these issues are fully discussed and referenced in the recent work of Kouhi (2012). As an example, Figures 8.3 and 8.4 show typical meshing for aortic valve using ANSYS and data extraction at t=0.07 sec and the contour of structural deformation with mesh coordinates that were recently developed at Swinburne University of technology (Kouhi, 2012).

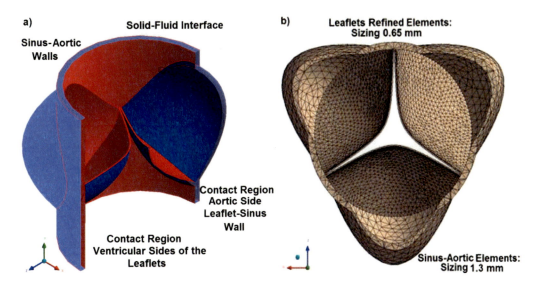

Figure 8.3. Solid Domain meshing strategy (a) Geometrical divisions and contact regions (b) element sizing.

Successively, it should be pointed out that the effect of the porosity on the fluid structure interaction for heart valve scaffolds has never been fully investigated. Such an investigation is critical in the area of tissue engineering, as there is a strong interplay between haemodynamic forces and porous scaffold structure interactions. Nevertheless, several macroscopic porous medium models based on Darcy's law or a modified Rayleigh-Darcy number are accessible in the literature (Sakamoto and Kulacki, 2007; Masoodi and Pillai, 2010). Moreover, in recent years, the lacking accuracy of Darcy's law at higher porosity and permeability has led to contemplation of non-Darcian models such as Forchhemier and Brinkman models. The Forchhemier model accounts for the additional flow resistance in terms of nonlinear drag owing to the higher flow velocities in the pores of solid matrix. Still, the viscous effect of no-slip condition near the solid boundaries is accounted for by the Brinkman model (Liu et al., 2007). However, other researchers have considered both non-Darcian terms together with inertial effect, which shown to be more pragmatic for most engineering applications (Hajipour and Dehkordi, 2011). Nevertheless, future research should be focused on the numerical analysis of the porous scaffold structure that is exposed to pulsating fluid forces and a constantly deforming solid structure.

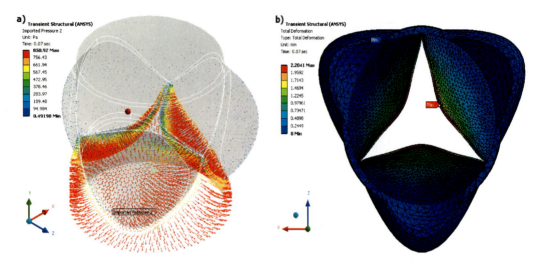

Figure 8.4. Transient FSI results at t=0.070 sec (a) imported pressure (b) total deformation: magnitudes and characteristic(c) Multi axial Von Mises stress (Kouhi, 2012).

8.3. Issues Related to Scaffold Selection

8.3.1. Decellularization Approach

As stated in Chapter 7, initially for tissue engineering development, xenograft or allografts were considered as an obvious choice as scaffold material because they are derived from the natural source. However, using such kind of biological materials as matrices has the potential to block the ingrowth of the cells. Hence, various decellularization techniques have been explored to remove the cellular components from such matrices. While doing so, there lies a chance of matrix properties being altered by the decellularization method used, thereby affecting the mechanical properties of the valve and its leaflets, which are highly undesirable. Hence, the future scope with the decellularization techniques lies in the development of a process that would not affect the mechanical structure of the valve and its leaflets (Brody and Pandit, 2007). The other approach, which recently enjoys tremendous popularity, is based on the design and optimization of polymer scaffold combined with natural materials. This approach relies mainly on the design and development of the correct scaffold made from the appropriate materials that is capable of mimicking the functionality of the valve and generating the autologous ECM; it is discussed in the next section.

8.3.2. Polymer and Natural Bio-Material Approach

With the polymer approach, the standard requirements for an ultimate scaffold template material for TE heart valve may include:

- From FDA approved implantable materials, i.e., to ensure compatibility with the recipient's tissue with no immunogenic or cytotoxic reactions.
- Retains surface characteristic that stimulates cell adherence and proliferation.
- Can be easily manufactured to correct shape of the valve and its leaflets.
- Must be easily sterilisable by validated methods.
- Provides a large surface area with effective degree of porosity and interconnectivity and micro-architecture with controllable rate of degradation.
- Permits attachment or incorporation of bioactive agents such as extra cellular matrix components, growth factors or peptides.

Therefore, the first challenge that any cell seeded scaffold template must meet is to ensure compatibility with the recipient's tissue as well as demonstrating efficacy as a replacement tissue. A TE construct should realistically approximate the physical properties of the tissue that is being replaced, i.e., shape, flexibility and texture, and it should be incorporated into the recipient's tissues without stimulating an inflammatory response. Therefore, with this approach, the selection of materials and manufacturing methods for the scaffold are interrelated and should be considered concurrently. These are discussed below.

8.3.2.1. Selection of Materials

The creation of suitable biomaterial for the valve scaffold requires the consideration of a number of factors such as the type of material (biological or synthetic or combination of both), surface topography and biocompatibility of materials, as these factors determine the success or failure of the tissue-engineering scaffold under in vitro and in vivo trials. Moreover, it is recognised that different applications demand different materials with specific properties and tissue-specific implants may be custom designed by combining various scaffold materials with different cell populations. Scaffold prototypes may be proposed with approved, biocompatible components that are biodegradable, bio-absorbable, and non-degradable or a combination of each depending on the intended anatomical function. Generally, in heart valve design, bovine pericardium is commonly proposed. The versatility of bovine pericardium is due to its characteristics of being durable and elastic and of possessing good suture retention strength.

However, significant selection criterion for constructing an ideal polymeric scaffold is the degree of porosity, interconnectivity, thickness, rate of degradation as well as the shape of scaffold. The production of materials tailored to such particular specifications can be achieved using synthetic materials, for example, biodegradable or non-biodegradable polymers such as polyglycolide (PGA), polycaprolactone (PCL), poly-L-lactide (PLLA) and polyurethane (PU). PU is also a widely used component in biomedical devices such as catheters and pacemakers. Although in general researchers have a good understanding of the acceptable degree of porosity and the range of scaffold mass and thickness required in the design of appropriate valve scaffold, still degradation in some polymers such PHO and PHA is quite lengthy. In fact, Sodian et al. reported that even after 24 weeks post-implantation of these scaffolds, some fragments were found (Sodian et al., 2000).

However, as pointed out previously, the advantage of using biological materials is that similar physical qualities of natural tissues can be maintained, and these materials provide natural environments that enhance physiological events such as cell attachment (Almany and Seliktar, 2005). On the other hand, biological materials do not offer the flexibility of custom-designed properties such as specific surface structure and/or pore size, which affect both the mechanical and cellular characteristics of biomaterials. Subsequently, currently there is a growing area of research into the production of biosynthetic and biomimetic materials that combines biological and synthetic components. A common approach that is being investigated widely is to cross-link biological scaffolds such as collagen and fibronectin to a synthetic element such as polyethylene glycol (PEG). Moreover, the cell attachment and migration may be improved by modifying the coating techniques, for example, polyurethane (PU) onto the pericardium. Studies *in vitro* have shown that surface topography affects cell attachment in that endothelial cells favourite smooth surfaces, whereas smooth muscle cells preferred coarse surfaces (Xu et al., 2004; Wong et al., 2006; Wong et al., 2010).

8.3.2.2. Issues Related to Manufacturing Methods

The design requirements need to be accurately defined and the manufacturing technology applied to produce the desired structural characteristics. As shown in Chapter 5, conventional techniques of scaffold template fabrication (e.g., fibre bonding, solvent casting and melt moulding) are inadequate for scaffold fabrication as they do not offer all of the required mechanical and surface characteristics to facilitate cell adherence and proliferation. The RP technologies including FDM, Stereolithography (SLA), Selective Laser Sintering (SLS), ink-jet-based systems, three-dimensional printing (3DP) and many other processes have several remarkable strengths and weaknesses when considered for TE applications. Constructs produced by SLA usually have a good surface finish and appearance, whereas those produced with 3DP or SLS have a sandy or diffuse texture due to the use of powders. For parts produced by SLA, a secondary process is required to completely cure the parts. Constructs made with 3DP and multi-jet modelling (MJM) can be very fragile and may not survive normal handling without post treatment. On the other hand, parts built by FDM have good strength immediately out of the machine and can be used for limited functional testing. The choice of method for scaffold template fabrication is, therefore, a question of setting priorities to determine the material and structural requirements critical to cell adhesion and tissue generation.

However, owing to the excellent ratio between the diameter and the length of the fibres (micro and nano scales) the electrospinning technique is widely proposed for tissue-engineering applications, which is capable of producing a wide range of polymers constructs (Dosunmu et al., 2006). The capability of the system is quite excellent, as the electro-spun fibre diameters could be varied from less than 100 nm to greater than 1000 nm. Such a variation is a function of a number of controllable parameters, including the type of polymers, and operating conditions, for example, solvent concentration and gap distance between the needle and the collector, etc. (Li and Xia, 2004). Moreover, recently, a novel manufacturing technique was introduced, which combines the FDM and Electrospinning methods (Chen et al., 2009). Such concept in principle offers a great potential for manufacturing complex and accurate scaffold configurations made from various biomaterials.

8.4. Issues Related to Cell Culture and Sources

Generally speaking, potential complications arise with the generation of tissue and maintaining survival of the cell after seeding it to the scaffold. In practice, large numbers of cells are needed to develop considerably small volume of the tissue. Thus, to generate large-size tissue, large volumes of cells are required. Basically, the problem of concern in this area is the limited availability of the donor sites. Moreover, there also arises question about the source from which the cells are stemmed. Mature cells require high amounts of oxygen but are low in potential for growth. Alternatively, immature cells, which are classified as embryonic cells, may result into significantly different cell growth (Leri et al., 2005).

With cell sources, in general, most researchers in the field followed the model of using vascular cells from the carotid artery and myofibroblasts. However, it is recognised now that proliferation and cell coverage of these cells are problematic and unrealizable to obtain for extensive clinical applications. Subsequently, additional cell sources have been considered by researchers for tissue engineering of heart valves including arterial fibroblasts, peripheral vein and stem cell, particularly human mesenchymal stem cells (HMSC cells)

Consequently, it is now well recognised that stem cells (embryonic, fetal, or adult sources, human and non-human), owing to their exceptional biological characteristics, have great potential for tissue-engineering applications and can be isolated from the somatic tissues (outer part of the body rather than organ). Moreover, a subgroup of cells exist within the human bone marrow that can be multiplied in the laboratory and stimulated to differentiate into multiple mesenchymal tissue cell types (MSCs), in both *in vitro* or *in vivo*, with excellent properties of malleability and self-renewal (Sutherland et al., 2005).

Sutherland et al. (2005) reported that the MSCs harvested from various species have the potential to regenerate an autologous valves from bone marrow-derived MSCs, together with a synthetic biodegradable scaffold (Sutherland et al., 2005).

8.5. Mechanical Simulation

It has been well documented that mechanical forces play a critical role in the homeostasis and remodelling of many load-bearing tissues and organs *in vivo* as well as in the stimulation of native and engineered tissue biosynthesis and development *in vitro* (Sodian et al., 2000; Morsi et al., 2007a). Mechanical simulation is essential in the generation of the appropriate tissues (Cacou et al., 2000; Mitchell et al., 2001), and as shown in Chapter 6, such concept has been applied in TEHV, where the correct physiological conditions have been simulated using specifically designed bioreactors (Hoerstrup et al., 2000(a); Hoerstrup et al., 2000(b); Rabkin et al., 2002).

Table 8.1. Current research issues related to the creation of TEHV

Manufacturing issues	Cells and Culture issues	HV Scaffold approaches	
		Decellularization	Polymer /and natural
Cell seeded scaffold template must be compatible with the recipient's tissue as well as demonstrating efficacy as a replacement tissue.	Other cell sources should be considered for future TEHV experiments, including stem cells.	Matrices, biological materials might stop the in-growth of the cells.	Biological and/or synthetic materials, surface topography and biocompatibility of materials.
The manufactured scaffold should approximate the physical properties of the tissue.	Potential problem ascends with the generation of tissue and preserving survival of the cell after seeding it to the scaffold.	Effects of decellularization method on mechanical properties of the leaflet should be considered.	Biosynthetic and biomimetic materials should be further investigated.
Technique for scaffold template fabrication is essential to determine the material and critical structural requirements to cell adhesion and tissue generation.	Optimal bioreactor parameters, including magnitude, duration, and rates of change of flows and pressures, must also be controlled.	Future decelluraization techniques not affecting the mechanical properties of leaflet are called for.	Porosity, thickness, and interconnectivity degradation rate, and shape are important selection parameters for choosing the ideal polymeric scaffold.
Production of nano-fibres by electrospinning polymeric materials for tissue engineering applications.			Resilient and flexible surfaces, leaflet designs are necessary for TEHVs to remodel with host.

However, with *in vitro* conditioning, the selected bioreactor parameters should include cycling hemodynamic forces, and the correct physiological conditions, of pressure shear stress etc. These testing should be carried out simultaneously with cell culture medium physical properties with growth factors etc for proper regeneration of ECM. For instance it was found that polymer-based TEHVs suffer from the deficiency of elastin in the ECM (Hoerstrup et al., 2000a) and an inadequate endothelial cell layer (Sodian et al., 2000), and such drawbacks could be attributed to the inadequate or inaccurtae mechanical and or chemical simulations.

Generally, it should be noted that the application of the forces generated by the bioreactor is gradually introduced to the scaffold to avoid "wash out of the cells" until the tissue reached certain maturity. However, with this type of *in vitro* conditioning, it is not clear yet which particular force factor is the predominate, as in these type of simulations all the hemodynamic forces, such as shear stress, pressure, tension, and flexure, are simultaneously generated; hence it is hard, if not impossible, to determine exactly the contribution from each mechanical force during this process of *in vitro* conditioning (Engelmayr, 2005).

Therefore, there is a need for the mechanical property and functionality of TEHVs to be thoroughly in vitro and in vivo tested. For example, our current understanding indicates that the effects of the growth factors, metalloproteinase, and their inhibitors could result in TEHVs with ECM generation and rate of degradation, which is similar to natural heart valves (Stock et al., 2000).

Table 8.1 illustrates some of the important issues that need to be further examined to enhance our current understanding of the concept of TEHV.

8.6. Closing Remarks

As highlighted in this chapter, in general, the basic approaches for the tissue-engineered heart valves that can be developed and designed depends fully on the scientific and engineering principles (Sacks et al., 2009). Moreover, what has been clearly illustrated here is that designing of robust tissue-engineering valve necessitates the optimization of various parameters, including the enduring durability of the valves, the correct shape and geometrical characteristics of it, the selection of the material and the degradation mechanism of the polymer components during tissue regeneration and the cellular reactions to all these factors, as well as the surrounding physiological conditions. These factors should be considered in terms of heart valve and the way it grows and alters the physiological demands. Moreover, to assist such development, a good understanding of the natural valves is necessary, and normally animals' models are used to gain this before human implantation.

It was also pointed out that the integrity of extracellular matrices (ECM) is the principle behind the durability of the heart valves. The ECM depends on how well the population of the cells is and the ability to overcome with different environments. Hence, the long-term success of the tissue-engineered valve depends on the capacity to maintain and remodel the ECM in vivo through mechanical signals and balance this with the stress transfer to the ECM, as a degradation rate is taking place. However, most of the studies reported in the literature have been conducted in vitro. Only a few in vivo studies that implanted either biological scaffold, in the form of decellularized heart valve or synthetic scaffold such as polyglycolic acid, polylactic acid and composites of polymer, have been conducted in animals or humans (Mendelson and Schoen, 2006). Another issue of concern is, since mechanical properties do not favour the placement in the aortic position to date, tissue engineering heart valves were implanted in the pulmonary position.

The fundamental issues of further research have been briefly highlighted here, which showed that the limitations of biological scaffolds include xenograft inciting inflammatory responses and the process of decellularization modifying the physical properties of valves. Moreover, although these limitations can be addressed using synthetic scaffolds, there are other challenges associated with synthetic scaffolds, such as designing optimal microstructures to support cell growth and achieving a balance between polymeric degradation and tissue formation. This polymer approach is quite popular and as a result, biomaterials research activities had increased throughout the world, over 20% per year in the last decade. The main aim of these activities is to develop a scaffold that possess good mechanical integrity and encourages cell growth and the reconstruction of ECM. We pointed out above that with this in mind, the exploration of hybrid materials and scaffolding materials could provide a pathway for the experimental studies and clinical applications of tissue engineering heart valves (Dong et al., 2008).

Moreover, utilizations of hemodynamic analysis and FSI to determine architecture of materials and characteristics could be invaluable tools in the optimization of the scaffold. This is normally followed by use of in vitro conditioning of the bioreactor that provides a controllable, self-regulating set up to quantify the hemodynamic forces and shear stress that should withstand during tissue conditioning. Moreover, it is now well recognised that stem cells, when differentiated properly, can be expected to provide a broader source of autologous cell lines for use in cardiovascular applications (Ivan, 2005). Hence, utilization of stem cells for tissue engineering of the heart would be an encouraging challenge for future development of TEHV.

In a nutshell, it is recognised that TEHV could effectively address valvular heart disease and to totally eliminate all the drawbacks related to herat valve repalcements and treatmentas discussed in Chapter 3. In addition, there is no doubt that a lot of research is still needed on both polymeric and decellularized biometrics tissue engineering approaches before TEHV can be sucessfuly implemented in human. In essence, tissue engineering can utilize the existing and future information and data in the field of biology, biochemistry and molecular biology to formulate the correct concept of generating new tissue. Likewise, progress in the science of bioengineering simulation and numerical modelling as well as advancement in biomaterials would, in principle, provide the right platform to engineer tissues of the required chracteristics, which will compliment the realistic application of engineering principles to human implementations. Equally, developments of genetic engineering, cloning and stem cell biology, which are dependent on each other, can further enhance our understaing of various tissue human diseases and greatly assist the development of correct treatments. (Lanza et al., 2007; Mol et al., 2004; Morsi et al., 2004).

References

Almany, L., and D. Seliktar, Biosynthetic hydrogel scaffolds made from fibrinogen and polyethylene glycol for 3D cell cultures, *Biomaterials*, 2005. 26(15): pp. 2467-2477.

Alperin, C., P.W. Zandstra and K. A. Woodhouse, Polyurethane films seeded with embryonic stem cell-derived cardiomyocytes for use in cardiac tissue engineering applications, *Biomaterials*, 2005. 26(35): pp. 7377-86.

Antonia, R. A., Conditional Sampling in Turbulence Measurement, *Annual Review of Fluid Mechanics*, 1982. 13(1): pp. 131-156.

Brody, S. and A. Pandit, Approaches to heart valve tissue engineering scaffold design, *J. Biomed. Mater Res. B Appl. Biomater.*, 2007. 83(1): pp. 16-43.

Butcher, J. T., A. M. Penrod, A. J. Garcia and R. M. Nerem, Unique morphology and focal adhesion development of valvular endothelial cells in static and fluid flow environments, *Arterioscler. Thromb. Vasc. Biol.* 2004. 24(8): pp. 1429-34.

Cacou, C., D. Palmer, D. A. Lee, D. L. Bader and J. C. Shelton, A system for monitoring the response of uniaxial strain on cell seeded collagen gels, *Medical Engineering and Physics*, 2000. 22(5): pp. 327-333.

Chandran, K., K. Burg and S. Shalaby, Soft Tissue Replacements, Biomedical Engineering Fundamentals, *CRC Press*, 2006. pp. 44-1-44-30.

Chen, R., Y. Morsi, S. Patel, Q. F. Ke and X. M. Mo, A novel approach via combination of electrospinning and FDM for tri-leaflet heart valve scaffold fabrication, *Frontiers of Materials Science in China*, 2009. 3(4), pp. 359-366.

Dreger, S. A., P. Thomas, E. Sachlos, A. H. Chester, J. T. Czernuszka, P. M. Taylor and M. H. Yacoub, Potential for synthesis and degradation of extracellular matrix proteins by valve interstitial cells seeded onto collagen scaffolds, *Tissue Eng*, 2006. 12(9): pp. 2533-40.

Dong N., J. Shi, P. Hu, S. Chen and H. Hong, Current progress on scaffolds of tissue engineering heart valves, *Frontiers of Medicine in China*, 2008. 2(3).

Dosunmu, O.O., Chase, G.G., Kataphinan, W., and Reneker, D.H., Electrospinning of polymer nanofibres from multiple jets on a porous tubular surface. *Nanotechnology*, 2006. 17(4): p. 1123-1127.

Engelmayr, J. G. C., Optimization of Engineered Heart Valve Tissueextracellular Matrix, Doctor of Philosophy, University of Pittsburgh, 2005.

Feugier, P., R. A. Black, J. A. Hunt and T. V. How, Attachment, morphology and adherence of human endothelial cells to vascular prosthesis materials under the action of shear stress, *Biomaterials*, 2005. 26(13), pp. 1457-66.

Hajipour, M. and Dehkordi, A.M., Mixed Convection in a Vertical Channel Containing Porous and Viscous Fluid Regions With Viscous Dissipation and Inertial Effects: A Perturbation Solution. *Journal of Heat Transfer*, 2011. 133(9): p. 092602.

Hoerstrup, S. P., R. Sodian, S. Daebritz, J. Wang, E. A. Bacha, D. P. Martin, A. M. Moran, K. J. Guleserian, J. S. Sperling, S. Kaushal, J. P. Vacanti, F. J. Schoen and J. E. Mayer, Jr., Functional living trileaflet heart valves grown in vitro, *Circulation*, 2000a. 102(19 Suppl 3), pp. III44-9.

Hoerstrup, S. P., R. Sodian, J. S. Sperling, J. P. Vacanti and J. E. Mayer, Jr., New pulsatile bioreactor for in vitro formation of tissue-engineered heart valves, *Tissue Eng.* 2000b.6(1): pp. 75-9.

Hollister, S. J., Porous scaffold design for tissue engineering, *Nature Materials,* 2005. 4(7): pp. 518-524.

Ivan, V., Heart Valve Tissue Engineering, *Circ. Res.* 2005. 97(8), pp. 743-755.

Kouhi, E., *An Advanced Fluid Structure Interaction Study of Tri-leaflet Aortic Heart Valve,* 2012. PhD Thesis, Swinburne University of Technology.

Labrosse, M. R., C. J. Beller, F. Robicsek and M. J. Thubrikar, Geometric modeling of functional trileaflet aortic valves: Development and clinical applications, *Journal of Biomechanics,* 2006. 39(14): pp. 2665-2672.

Lanza, R. P., R. S. Langer and J. P. Vacanti, *Principles of tissue engineering:* Elsevier Academic Press, 2007.

Layland, S. K., F. Optiz, M. Gross, C. Doring, K. J. Halbhuber, F. Schirrmeister, T. Wahlers and U. A. Stock, Complete dynamic repopulation of decellularized heart valves by application of defined physical signals—an in vitro study, *Cardiovasc. Res.,* 2003. 6D(3): pp. 497-509.

Leri, A., J. Kajstura and P. Anversa, Cardiac Stem Cells and Mechanisms of Myocardial Regeneration, *Physiological Reviews,* 2005. 85(4): pp. 1373-1416.

Leveque, R. J. and Z. Li, Immersed interface method for elliptic equations with discontinuous coefficients and singular sources, *SIAM Journal on Numerical Analysis,* 1994. 31(4): pp. 1019-1044.

Li, D., and Y. Xia, Electrospinning of nanofibres: Reinventing the wheel?, *Advanced Materials,* 2004. 16(14): pp. 1151-1170.

Liechty, K. W., T. C. MacKenzie,A. F. Shaaban, A. Radu, A. M. Moseley, R. Deans, D. R. Marshak and A. W. Flake, Human mesenchymal stem cells engraft and demonstrate site-specific differentiation after in utero transplantation in sheep, *Nat. Med.* 2000. 6(11): pp. 1282-6.

Liu, H., Patil, P.R., and Narusawa, U., On Darcy-Brinkman equation: Viscous flow between two parallel plates packed with regular square arrays of cylinders. *Entropy.* 2007. 9(3): p. 118-131.

Masoodi, R. and Pillai, K.M., Darcy's law-based model for wicking in paper-like swelling porous media. *AIChE Journal.* 2010. 56(9): p. 2257-2267.

Mendelson, K., and F. J. Schoen, Heart valve tissue engineering: concepts, approaches, progress, and challenges, *Ann. Biomed. Eng.* 2006. 34(12): pp. 1799-819.

Mol, A., C. V. Bouten, P.T. Frank F. P. Baaijens, G. Zund, I. M. Turina and S. P. Hoerstrup, Tissue engineering of semilunar heart valves: Current status and future development, *Journal of heart valve disease.* 2004.13(2): pp. 272-80.

Mercer, J. L., M. Benedicty and H. T. Bahnson, The geometry and construction of the aortic leaflet, *J. Thorac Cardiovasc. Surg.* 1973. 65(4), pp. 511-8.

Meuris, B., Calcification of Aortic Wall Tissue in Prothetic Heart Valves: Initation, *Influencing Factors and Strategies Towards Prvention,* [dissertation], Leuven University, 2007.

Mitchell, S. B., J. E. Sanders, J.L. Garbini and P. K. Schuessler, A device to apply user-specified strains to biomaterials in culture, *IEEE Transactions on Biomedical Engineering,* 2001. 48(2): pp. 268-273.

Morsi, Y. S., Birchall I. E. and F. L. Rosenfeldt, Artificial aortic valves: an overview, *Int. J. Artif. Organs.,* 2004. 27(6): pp. 445-51.

Morsi, Y. S., W. W. Yang, A. Owida and C. S. Wong, Development of a novel pulsatile bioreactor for tissue culture, *Journal of Artificial Organs,* 2007a. 10(2): pp. 109-114.

Morsi, Y. S., W. W. Yang, C. S. Wong and S. Das, Transient fluid–structure coupling for simulation of a trileaflet heart valve using weak coupling, *Journal of Artificial Organs,* 2007b. 10(2): pp. 96-103.

Morsi, Y. S, P.S., Zhang Li, Advancement of lung tissue engineering: an overview. *Int. J. Biomedical Engineering and Technology,* 2011. 5: p. 195-210.

Norton, T. and D. W. Sun, Computational fluid dynamics (CFD)—an effective and efficient design and analysis tool for the food industry: *A review, Trends in Food Science & Technology,* 2006. 17(11): pp. 600-620.

Rabkin, E., S. P. Hoerstrup, M. Aikawa, J. E. Mayer Jr. and F. J. Schoen, Evolution of cell phenotype and extracellular matrix in tissue-engineered heart valves during in-vitro maturation and in-vivo remodeling, *Journal of Heart Valve Disease,* 2002. 11(3): pp. 308-314.

Ramaswamy, S., D. Gottlieb, G. C. Engelmayr, Jr., E. Aikawa, D. E. Schmidt, D. M. G. Leon, V. L. Sales, J. E. Mayer, Jr. and M. S. Sacks, The role of organ level conditioning on the promotion of engineered heart valve tissue development in-vitro using mesenchymal stem cells, *Biomaterials,* 2010. 31(6): pp. 1114-25.

Sacks, M. S., F. J. Schoen and J. E. Mayer, *Bioengineering Challenges for Heart Valve Tissue Engineering, Annual Review of Biomedical Engineering,* 2009. 11(1): pp. 289-313.

Sakamoto, H. and Kulacki, F.A., Buoyancy driven flow in saturated porous media. *Journal of Heat Transfer,* 2007. 129(6): p. 727-734.

Schoen, F. and R. Levy, Tissue heart valves: Current Challenges and future research perspectives, *Founders Aw 25th Annual Meeting Soc. Biomaterials.* 1999. Pp: 441-65.

Simon, P., M. T. Kasimiri, G. Seebacher, G. Weigel, R. Ullrich, U. Salzer-Muhar, E. Rieder, and E. Wolner, Early failure of the tissue-engineered porcine heart valve SYNERGRAFT™ in pediatric patients, *Eur. J. Cardiothorac. Surg.,* 2003. 23: pp. 1002-1006.

Sodian, R., S. P. Hoerstrup, J. S. Sperling, S. Daebritz, D. P. Martin, A. M. Moran, B. S. Kim, F. J. Schoen, J. P. Vacanti and J. E. Mayer, Jr., Early in vivo experience with tissue-engineered trileaflet heart valves, *Circulation,* 2000. 102(19): pp. III22-III29.

Stamm, C., A. Khosravi, N. Grabow, K. Schmohl, N. Treckmann, A. Drechsel,M. Nan, K. P. Schmitz, A. Haubold and G. Steinhoff, Biomatrix/polymer composite material for heart valve tissue engineering, *Ann. Thorac. Surg.* 2004. 78(6): pp. 2084-93.

Stock, U. A., M. Nagashima, P. N. Khalil, G. D. Nollert, T. Herden, J. S. Sperling, A. Moran,J. Lien, D. P. Martin, F. J. Schoen,J. P. Vacanti and J. E. Mayer, J. E., Jr., Tissue-Engineered valved conduits in the pulmonary circulation, *J. Thorac Cardiovasc. Surg.* 2000. 119(4): pp. 732-740.

Sutherland, F. W. H., T. E. Perry,Y. Yu, M. C. Sherwood, E. Rabkin, Y. Masuda, G. A. Garcia, D. L. McLellan, G. C. Engelmayr, Jr., M. S. Sacks, F. J. Schoen and J. E. Mayer, Jr., From Stem Cells to Viable Autologous Semilunar Heart Valve, *Circulation,* 2005. 111(21): pp. 2783-2791.

SWANSON, W. M., and R. E. CLARK, Dimensions and Geometric Relationships of the Human Aortic Value as a Function of Pressure, *Circ. Res.* 1974. 35(6): pp. 871-882.

Terada, S., M. Sato, A. Sevy and J. P. Vacanti, Tissue Engineering in the Twenty-First Century, *Yonsei Medical Journal,* 2000. 41(6): pp. 685-691.

Thubrikar, M.J., Skinner, J.R., Eppink, R.T., and Nolan, S.P., Stress analysis of porcine bioprosthetic heart valves in vivo. *Journal of Biomedical Materials Research.* 1982. 16(6): p. 811-826.

Van Mook, W. N. K. A., and L. Peeters, Severe cardiac disease in pregnancy, part I: hemodynamic changes and complaints during pregnancy, and general management of cardiac disease in pregnancy, *Current Opinion in Critical Care.* 2005.11(5): pp. 430-434.

Wong, CS, Sgarioto, M Owida AA, Yang W, Rosenfeldt FL and Y.S. Morsi, Polyethyleneterephthalate Provides Superior Retention of Endothelial Cells During Shear Stress Compared to Polytetrafluoroethylene and Pericardium. *Heart Lung and Circulation,* 15(6): p. 371-377, 2006.

Wong CS, Patel SS, Chen R, Owida A and Y.S. Morsi, Biomimetic Electrospun Gelatin-Chitosan Polyurethane for Heart Valve Leaflets. *Journal of Mechanics in Medicine and Biology.* 2010. 0(1): p. 1-26

Xu, C., F. Yang,S. Wang and S. Ramakrishna, In vitro study of human vascular endothelial cell function on materials with various surface roughness, *J. Biomed. Mater Res.* 2004. 71A(1): pp. 154-61.

Zund, G., C. K. Breuer, T. Shinoka, P. X. Ma, R. Langer, J. E. Mayer and J. P. Vacanti, The in vitro construction of a tissue engineered bioprosthetic heart valve, *European Journal of Cardio-thoracic Surgery.* 1997. 11(3): pp. 493-497.

Index

A

abnormal valve motion, 2, 162
accessibility, 162
acclimatization, 2
acetic acid, 56, 94
acid, 44, 46, 47, 48, 50, 52, 53, 54, 55, 57, 60, 61, 62, 64, 65, 68, 69, 70, 72, 73, 74, 75, 89, 121, 122, 124, 128, 130, 135, 136, 141, 142, 150, 154, 156, 173
acidic, 49, 59
acrylate, 61
adaptability, 141
adaptation, 96
additives, 68, 137
adhesion, 3, 4, 44, 45, 51, 58, 63, 65, 107, 122, 131, 136, 141, 142, 156, 157, 164, 170, 172, 175
adhesions, 67
adhesives, 66
adsorption, 74, 136, 141
advancement, 36, 68, 162, 174
advancements, 4
adverse effects, 99, 149
AFM, 72
age, 1, 7, 8, 19, 35, 162
ageing population, 7
aging population, 1, 162
alanine, 74
alcohols, 55, 58
algae, 47
aluminium, 145
American Heart Association, 5, 23
amines, 58
amino, 45, 46, 59, 60, 62, 65, 70, 136, 149
amino acid(s), 45, 46, 59, 60, 62, 65, 70
ammonium, 78
anatomy, 8
anesthetics, 64

anisotropy, 154
ankles, 15
anticoagulant, 30, 35, 162
anticoagulation, 28, 30, 37, 38, 39, 162
antigenicity, 46, 125
antiphospholipid syndrome, 40
antisense, 69
aorta, 9, 10, 11, 13, 14, 17, 27, 28, 30, 33, 37, 135, 137, 144, 151, 164
aortic insufficiency, 30, 39
aortic regurgitation, 17, 32, 38
aortic stenosis, 16, 17, 18, 19, 22
aqueous solutions, 50
arterioles, 105
artery(s), 10, 11, 12, 14, 20, 32, 51, 105, 115, 137, 140, 141, 144, 145, 146, 151, 171
arthropods, 49
articular cartilage, 63
artificial heart valve, 2, 25, 26, 27, 32, 34, 36, 39, 43, 161, 163
ascorbic acid, 58
assessment, 17, 21
asymptomatic, 18
atherosclerosis, 18
atoms, 21, 61
atria, 9, 10, 13, 14
atrioventricular node, 14
atrium, 8, 10, 13, 14, 106
attachment, 67, 68, 77, 126, 127, 129, 130, 132, 133, 136, 137, 145, 146, 147, 149, 150, 151, 164, 169, 170
auscultation, 20
autolysis, 149
awareness, 62

B

background information, 5

bacteria, 1, 15
bacterial fermentation, 48, 51
bacterial infection, 57
bacteriostatic, 51
barium, 69
barriers, 2
base, 55, 83, 84, 85, 86, 104, 162, 163
basement membrane, 126, 131
batteries, 66
BD, 140
bending, 12, 137, 164
benefits, 68
bicarbonate, 78
bicuspid, 9, 16, 18, 19
bioactive agents, 63, 66, 169
biochemistry, 174
biocompatibility, 4, 46, 47, 49, 50, 51, 54, 56, 58, 60, 62, 64, 65, 66, 68, 74, 75, 81, 95, 126, 149, 157, 162, 169, 172
biocompatibility test, 75
biodegradability, 45, 51, 62, 64, 68, 74
biodegradable materials, 52, 58, 119
biodegradation, 51, 54, 60, 68
biologic tissue, 2
biological responses, 94
biomaterials, 3, 4, 59, 60, 61, 62, 69, 72, 116, 120, 128, 141, 144, 148, 149, 150, 155, 159, 163, 170, 174, 176
biomedical applications, 58, 60, 75
biopolymer(s), 47, 48, 49, 64, 138
biosynthesis, 70, 171
blends, 96
blood, 1, 7, 8, 9, 10, 11, 12, 13, 14, 15, 16, 19, 20, 21, 25, 26, 28, 32, 35, 36, 37, 38, 41, 45, 46, 51, 63, 114, 115, 137, 138, 144, 148, 149, 166
blood circulation, 15
blood clot, 15, 28, 63
blood flow, 1, 2, 8, 9, 14, 15, 17, 20, 25, 26, 32, 36, 38, 137, 166
blood pressure, 14
blood stream, 15
blood vessels, 15, 20, 21, 45, 114, 115
bonding, 58, 60, 61, 65, 78, 79, 80, 83, 93, 170
bonds, 48, 56, 59, 60, 62, 65, 122
bone(s), 17, 43, 45, 46, 51, 52, 53, 54, 56, 57, 58, 64, 69, 71, 82, 83, 90, 92, 93, 94, 95, 96, 100, 115, 116, 117, 142, 144, 146, 148, 151, 152, 153, 155, 156, 157, 171
bone cells, 94, 95
bone growth, 57
bone marrow, 142, 144, 146, 148, 151, 152, 153, 157, 171
brain, 15

breathing, 15, 20
breathlessness, 15
Butcher, 161, 175
by-products, 4, 54, 74

C

CAD, 83, 85, 164, 165
calcification(s), 1, 8, 17, 18, 36, 43, 44, 119, 123, 125, 129, 130, 159, 162
calcium, 10, 16, 36, 84
campaigns, 110
candidates, 52, 57
capillary, 90, 140
carbohydrates, 46
carbon, 1, 8, 27, 28, 37, 54, 62, 78, 82, 92, 95, 161
carbon dioxide, 8, 78, 82
carbon nanotubes, 92
carboxylic acid, 55
carcinoma, 117
cardiac muscle, 14
cardiac pacemaker, 58, 65
cardiac surgery, 34, 42
cardiopulmonary bypass, 27
cardiovascular disease, 7
cardiovascular function, 25
cardiovascular physiology, 7
cardiovascular system, 43
carotid arteries, 140
cartilage, 43, 45, 51, 83, 90, 96, 104, 115, 136, 159
casting, 53, 64, 78, 81, 93, 138, 146, 170
catabolites, 101, 114
catalyst, 56, 63
cell biology, 174
cell culture, 48, 90, 98, 99, 102, 114, 126, 134, 135, 162, 163, 173, 175
cell death, 134
cell differentiation, 114, 119, 148
cell line(s), 117, 125, 144, 174
cell membranes, 104
cell surface, 134
cellulose, 49, 51
ceramic, 88, 95
challenges, 5, 6, 23, 33, 41, 57, 72, 157, 161, 174, 176
charge density, 91
chemical, 17, 30, 36, 38, 44, 45, 46, 48, 49, 50, 51, 52, 53, 55, 56, 61, 68, 71, 83, 97, 100, 121, 122, 125, 126, 127, 135, 136, 148
chemical properties, 45, 46, 136
chemical structures, 49
chemicals, 4, 122, 124
childhood, 17

Index

181

children, v, 19, 30, 31, 36, 43, 162
China, 154, 175
chitin, 45, 49, 71, 73, 74, 136
chitosan, 44, 45, 49, 50, 51, 69, 71, 73, 74, 75, 92, 94, 95, 96, 127, 128, 132, 136, 139, 141, 146, 149, 154, 157, 159, 178
chloroform, 80, 83
cholesterol, 18, 51
chorionic villi, 144, 152
circulation, 8, 111
classes, 59, 100
classification, 10, 32, 47
cleavage, 59
clinical application, 3, 52, 171, 174, 176
cloning, 174
closure, 12, 17, 143, 145, 146, 166
clustering, 103
CO_2, 78, 96, 155
coatings, 136
cobalt, 32
collagen, 12, 13, 45, 46, 47, 51, 69, 71, 94, 121, 122, 123, 124, 126, 128, 129, 130, 131, 134, 135, 136, 138, 139, 140, 142, 145, 146, 149, 150, 151, 152, 153, 158, 159, 170, 175
commercial, 61, 72, 123, 165
commissure, 33
commodity, 54
compatibility, 27, 30, 31, 62, 63, 148, 154, 169
complexity, 8, 36, 91, 166
compliance, 2, 106, 107, 111, 118
complications, 2, 30, 32, 37, 38, 119, 161, 171
composites, 71, 74, 135, 174
composition, 17, 50, 63, 74, 75, 126, 136, 139, 141, 147, 158
compression, 27, 53, 113
computed tomography, 23, 143
computer, 111, 115
conciliation, 26
condensation, 52, 56, 57, 58, 62
conditioning, 3, 4, 5, 97, 98, 99, 100, 104, 109, 111, 114, 115, 119, 122, 123, 129, 130, 140, 145, 151, 152, 162, 173, 174, 177
conductivity, 65
configuration, 32, 97, 120, 163, 164
congenital malformations, 8
congestive heart failure, 15
connective tissue, 8, 17, 51
connectivity, 81, 82, 88, 126
consensus, 115
constituents, 3
construction, 34, 44, 57, 85, 104, 150, 160, 176, 178
consumption, 28
containers, 107, 135

contamination, 93, 97, 112, 113
contour, 167
controversial, 31
cooling, 84
copolymerisation, 55, 71
copolymerization, 139
copolymer(s), 36, 52, 54, 55, 68, 70, 73, 136, 137, 142, 145, 150
coronary arteries, 17
coronary artery bypass graft, 20
coronary artery disease, 7
coronary heart disease, 7
correlation, 2
cortical bone, 57
cosmetic, 45, 51, 88
cost, 4, 25, 38, 64, 83, 85, 90, 96, 100
CPC, 125, 130
cross-linking reaction, 55
crystalline, 49, 52, 53, 55, 60
crystallinity, 52, 53, 54, 64
crystallization, 70
crystals, 80, 81, 121
CT, 20, 21, 22, 143, 165
cues, 141
cultivation, 4, 101, 104, 105, 126
culture, 4, 52, 97, 99, 100, 101, 102, 103, 105, 107, 108, 109, 110, 111, 112, 113, 114, 115, 116, 117, 124, 130, 134, 135, 136, 139, 141, 146, 148, 151, 152, 153, 176, 177
culture media, 99
culture medium, 97, 101, 108, 109, 110, 111, 114, 115
cure, 170
CVD, 7
cycles, 11, 14, 28, 36
cycling, 173
cytotoxicity, 46, 48, 51

D

damages, 28, 121
deacetylation, 49
death rate, 7
deaths, 7
defects, 19, 35, 43, 57, 63, 64, 93, 117, 153
deformation, 4, 12, 27, 63, 124, 130, 137, 141, 150, 151, 164, 166, 168
degradation, 2, 36, 44, 49, 50, 52, 53, 54, 55, 56, 57, 59, 60, 61, 64, 65, 66, 67, 68, 70, 72, 73, 74, 75, 95, 119, 120, 127, 128, 131, 132, 135, 136, 137, 138, 139, 141, 145, 146, 147, 149, 150, 151, 152, 162, 169, 172, 173, 174, 175
degradation mechanism, 53, 56, 173

182 Index

degradation process, 53, 55
degradation rate, 44, 52, 57, 59, 61, 64, 65, 120, 128, 137, 141, 147, 151, 172, 173
dehydration, 46, 56
dehydrochlorination, 56
deposition, 18, 36, 83, 93, 96, 130, 145
deposits, 16, 18
deprivation, 45
derivatives, 60, 61
destruction, 149, 162
detectable, 125, 145, 153
detection, 20, 21
detergents, 121, 124, 125, 130, 131, 132, 148
developing countries, 7, 25
diaphragm, 104, 107, 112
diastole, 7, 11, 12, 14, 147
dielectric relaxations, 124
diffusion, 53, 80, 99, 136, 139, 149, 164
diffusivity, 91, 167
digestion, 125
diisocyanates, 58, 65
diluent, 55
direct observation, 165
disaster, 123
discs, 29, 37, 104
diseases, 1, 5, 7, 8, 10, 15, 16, 17, 20, 25, 36, 67
disorder, 121, 166
displacement, 20
distillation, 47
distilled water, 92
distribution, 46, 48, 98, 103, 114, 134, 151
DNA, 125, 133, 134, 138, 140, 142
DNase, 124, 127, 129, 130, 131, 132, 133
donors, 30, 148
drawing, 32
drug carriers, 75
drug delivery, 47, 48, 52, 55, 56, 57, 58, 59, 61, 62, 64, 65, 66, 69, 71, 74, 75, 94
drug release, 59, 74
drugs, 59, 64
drying, 47, 78, 80, 123, 129, 149
DSC, 123
durability, 2, 26, 42, 161, 162, 173

E

EAE, 22
ECM, 3, 4, 5, 11, 12, 13, 43, 45, 51, 67, 97, 104, 115, 119, 120, 121, 122, 123, 124, 125, 126, 127, 128, 129, 130, 131, 132, 133, 136, 138, 139, 140, 148, 150, 151, 152, 163, 168, 173, 174
ECs, 127, 132, 149
elaboration, 139

elastic deformation, 27
elastin, 12, 13, 122, 123, 126, 128, 129, 130, 131, 134, 136, 138, 139, 140, 142, 152, 153, 160, 173
elastomers, 58, 70
electric field, 90
electron microscopy, 141
electron(s), 21, 141
electrospinning, 75, 77, 80, 83, 87, 90, 91, 92, 93, 96, 127, 142, 154, 175
emboli, 15, 28, 37
embolization, 28, 37
embryonic stem cells, 148
encapsulation, 69, 124
endocarditis, 1, 8, 10, 15, 32, 36, 38
endocardium, 1, 15
endothelial cells (ECs), 12, 13, 114, 120, 123, 124, 125, 126, 127, 129, 130, 131, 134, 137, 138, 140, 144, 145, 148, 149, 151, 152, 154, 160, 170, 175
endothelium, 125
endurance, 25, 139
energy, 12, 86, 88
engineering, ix, 3, 4, 5, 6, 23, 41, 43, 45, 51, 52, 65, 66, 68, 72, 77, 83, 88, 89, 90, 92, 94, 95, 97, 98, 104, 117, 119, 120, 136, 137, 144, 145, 153, 154, 155, 157, 159, 160, 161, 162, 163, 168, 173, 174, 176
enlargement, 17, 21, 166
environment(s), 2, 4, 32, 43, 61, 65, 97, 99, 100, 104, 105, 114, 115, 117, 126, 149, 167, 170, 173, 175
environmental factors, 2
enzyme(s), 45, 54, 98, 128
equipment, 84, 90
erosion, 53, 56, 59, 65, 70, 141, 145
ester, 52, 53, 54, 58, 59, 60, 62, 65, 66, 68, 69, 74, 75, 142
ester bonds, 52, 54, 62, 65, 66
ethylene, 72, 94, 136, 141, 146
ethylene glycol, 72, 136, 141, 146
ethylene oxide, 94
Europe, 123
evaporation, 80, 90
evidence, 124, 127, 140
evolution, 40, 42, 158
exercise, 34
exertion, ix, 161
exoskeleton, 49
exposure, 122
Extra Cellular Matrix (ECM), 3, 4, 5, 11, 13, 43, 140
extracellular matrix, 48, 105, 116, 119, 123, 125, 126, 134, 137, 138, 141, 144, 148, 149, 157, 158, 161, 175, 177
extraction, 48, 80, 129, 167
extrusion, 53

Index

F

fabrication, 2, 4, 42, 53, 69, 71, 72, 77, 78, 80, 81, 82, 83, 87, 88, 93, 94, 95, 96, 135, 150, 154, 155, 158, 159, 170, 172, 175
fainting, 15, 20
fascia, 31, 45
FDA, 53, 54, 128, 137, 169
fear, 67
feedstock, 91
FEM, 69
fever, 7, 15
fiber(s), 12, 50, 72, 74, 78, 80, 90, 92, 96, 126, 129, 131, 134, 135, 142, 145, 146, 149, 150, 151
fiber bundles, 135
fiber networks, 142
fibrin, 128, 134, 135, 139, 142, 143, 149
fibrinogen, 135, 149, 175
fibrinolysis, 149
fibroblast growth factor, 127, 152
fibroblasts, 12, 13, 120, 124, 135, 137, 138, 140, 141, 144, 145, 146, 152, 153, 154, 157, 160, 171
fibrous cap, 56
fibrous tissue, 124, 140
filament, 90
films, 50, 75, 136, 139, 150, 175
finite element method, 115
first generation, 100, 101
fixation, 2, 32, 36, 52, 54, 149, 162
flex, 116, 142, 146
flexibility, 9, 27, 50, 65, 141, 149, 150, 151, 169, 170
flocculation, 50
fluctuations, 9, 104
fluid, 6, 15, 99, 103, 105, 164, 165, 166, 167, 175, 177
foams, 53, 64
food, 117, 177
Food and Drug Administration, 53, 54, 128
food industry, 177
force, 103, 104, 135, 173
Ford, 5, 23
formaldehyde, 61
formation, 4, 15, 28, 35, 37, 45, 54, 56, 65, 90, 91, 102, 109, 112, 116, 129, 138, 140, 141, 142, 148, 151, 152, 153, 174, 176
fragments, 124, 136, 170
freedom, 35
freezing, 121, 122, 148, 164
fuel cell, 66
fungi, 15, 49

G

gangrene, 15
gel, 135, 138, 139, 142, 143, 149, 155, 160
gene expression, 67
genes, 45
genetic engineering, 174
geometry, 4, 85, 91, 93, 115, 117, 135, 138, 141, 164, 165, 176
glass transition, 52, 55, 59
glass transition temperature, 52, 55, 59
glucose, 58, 65
glue, 60
glycine, 45, 47, 73
glycol, 52, 64, 68, 93, 170, 175
glycosaminoglycans, 45, 134, 140
google, 42, 70
gravity, 46
grid generation, 165
grouping, 15
growth factor, 63, 78, 97, 134, 149, 152, 169, 173
growth rate, 80, 115
guidance, 147

H

hard tissues, 45
hardness, 124
healing, 43, 49, 63
health, 1, 8, 21, 22, 25, 35
heart disease, 1, 22
heart failure, 7, 15, 16
height, 110
hemorrhage, 2, 161
hexane, 57
histology, 13, 142
history, 52, 78
homeostasis, 171
homogeneity, 26
homograft prosthesis implantation, 30
hormones, 152, 153
host, 2, 31, 67, 68, 115, 125, 137, 162, 172
housing, 28, 35, 37
human body, 53, 54, 97, 114
humidity, 91
hyaluronic acid (HA), 48, 136
hybrid, 4, 61, 66, 93, 125, 127, 174
hydrocortisone, 152
hydrogels, 51, 58, 63, 65, 66, 71, 72, 73, 134, 136, 139, 150, 158
hydrogen, 46, 58, 60
hydrogen bonds, 46

Index

hydrolysis, 46, 47, 50, 52, 53, 54, 59, 62, 64, 65, 66, 68, 72, 122
hydrophilicity, 68, 137
hydrophobicity, 53, 68, 92, 137, 141
hydroxyapatite, 85, 88, 96, 157
hydroxyl, 52, 136, 149
hypertension, 17
hypertrophy, 19

I

ID, 95
ideal, 8, 17, 26, 34, 57, 88, 99, 115, 128, 135, 141, 144, 149, 164, 169, 172
idiopathic, 18
image(s), 20, 21, 143
immersion, 94
immune reaction, 45
immune response, 30, 44, 67, 122, 127, 148
immunofluorescence, 134, 138
immunogenicity, 3, 59, 65, 68, 120, 126, 160
implants, 53, 54, 57, 58, 65, 93, 101, 169
improvements, 4, 109
in utero, 176
incidence, 1, 30
incubator, 109, 110
inefficiency, 1
infants, 43
infection, 1, 15, 30, 36, 38, 43, 162
inferior vena cava, 13
inflammation, 1, 149, 163
inflammatory responses, 174
ingredients, 44
inguinal, 70
inguinal hernia, 70
inhibitor, 124, 129, 149
injury, 1, 63
insulin, 152
integration, 57, 115, 135
integrity, 2, 9, 27, 36, 37, 58, 59, 68, 121, 126, 128, 131, 134, 136, 137, 142, 143, 162, 164, 173, 174
interface, 135, 176
interference, 15
internal environment, 115
internal fixation, 53, 64
intervention, 22
ionic solutions, 122, 148
ions, 122
irradiation, 95, 138
isolation, 106
issues, 3, 4, 69, 98, 137, 163, 164, 166, 167, 172, 173, 174
Italy, 72

J

Japan, 71
joints, 17, 51
Jordan, 5, 116

K

keratin, 45
kinetics, 44, 73
Krebs cycle, 55

L

lack of control, 44, 65
lactate dehydrogenase, 134
lactic acid, 52, 53, 54, 64, 73, 95
Lagrangian formulation, 167
laminar, 13, 166
lamination, 78, 83
layering, 144
leaching, 53, 64, 78, 79, 81, 93, 146, 150
lead, 4, 8, 11, 15, 16, 28, 32, 61, 66, 68, 149, 164
leakage, 32, 35
left atrium, 9, 10
left ventricle, 9, 10, 11, 13, 16, 17, 105
legs, 15
lens, 66
lesions, 162
lifetime, 162
ligament, 46
ligand, 67
light, 57, 92, 124, 150
liquid phase, 80, 86, 93
liquids, 51, 91
lithium, 66
lithography, 143, 147
liver, 83, 102, 136
local stress, 1
longevity, 30, 32, 38
low risk, 10, 30, 38
lymph, 75
lymph node, 75
lysine, 135, 149
lysis, 121, 122, 126

M

macromolecules, 63, 92
magnetic field, 20
magnitude, 17, 35, 99, 128, 166, 172

majority, 26, 34, 125
mammalian cells, 103
mammalian tissues, 99
mammals, 45
man, 44, 50
management, 22, 40, 178
manufacturing, 4, 5, 6, 27, 44, 48, 63, 68, 77, 78, 81, 83, 85, 87, 93, 95, 98, 128, 141, 163, 164, 169, 170
Marfan syndrome, 17
marrow, 146, 148, 151, 157
Mars, 176
mass, 34, 35, 45, 59, 98, 99, 101, 102, 103, 105, 114, 117, 136, 137, 141, 145, 164, 169
mass loss, 136
matrix, 12, 43, 67, 92, 121, 122, 124, 125, 126, 129, 134, 135, 136, 138, 139, 140, 141, 145, 149, 150, 153, 155, 156, 157, 158, 159, 160, 168, 169
matrix metalloproteinase, 153
matter, 30, 31
measurement, 23
measurements, 2, 134, 165
mechanical performances, 59
mechanical properties, 3, 4, 27, 44, 50, 53, 54, 56, 58, 59, 60, 64, 65, 67, 74, 83, 87, 90, 92, 93, 98, 104, 115, 122, 124, 125, 127, 130, 137, 141, 142, 145, 146, 147, 149, 150, 151, 168, 172, 174
mechanical stress, 2, 36, 104, 115, 137, 164
media, 48, 50, 74, 100, 110
medical, 40, 41, 51, 52, 53, 54, 58, 59, 63, 65, 67, 70, 81, 109, 154
medicine, 73
melt, 56, 57, 80, 170
melting, 52, 55, 80, 85, 89
melting temperature, 55, 85, 89
melts, 86
membranes, 2, 10, 35, 36, 50, 66, 83, 92, 121, 124, 125
mesenchymal stem cells, 127, 142, 144, 152, 155, 171, 176, 177
metabolism, 36
metabolites, 59, 101, 114
metalloproteinase, 173
methylene chloride, 78, 80
microgravity, 103, 114, 117
microorganisms, 15
microscopy, 124, 141, 142
microspheres, 74, 94
microstructure(s), 23, 82, 135, 141, 164, 174
migration, 93, 134, 170
mineralisation, 2, 35
Minneapolis, 33
mitral valve, 8, 9, 10, 16, 17, 20, 21, 22

mitral valve prolapse, 8
mixing, 59, 92, 103, 114
MMPs, 153
modelling, 93, 158, 162, 164, 165, 166, 170, 174
models, 73, 83, 89, 126, 140, 165, 167, 173
modifications, 27, 28, 29, 36, 37, 83
modules, 165
modulus, 53, 54, 128
mold(s), 143
molecular biology, 174
molecular structure, 68
molecular weight, 46, 54, 55, 56, 62, 74
molecular weight distribution, 74
molecules, 50, 78, 128, 136, 158
momentum, 8
monolayer, 124, 126, 131, 140
monomers, 52, 57, 59, 60, 61
morphology, 54, 67, 68, 69, 92, 120, 122, 123, 126, 127, 128, 136, 139, 141, 142, 146, 149, 151, 175
mortality, 7, 162
moulding, 32, 53, 135, 138, 149, 155, 164, 170
MRI, 20, 21, 23
multifunctional monomers, 57
multiplier, 167
murmur, 21
muscles, 9, 10, 11, 33, 45
muscular tissue, 123
musculoskeletal, 54, 64, 153
myocardial infarction, 7
myofibroblasts, 12, 13, 120, 124, 125, 126, 130, 134, 135, 138, 139, 140, 142, 144, 153, 158, 171

N

NaCl, 124, 130
nanofabrication, 77, 83
nanofibers, 75, 96, 154, 157, 176
nanomaterials, 71
nanoparticles, 61, 74, 75, 127, 133, 154
nanotube, 95
natural polymers, 67
necrosis, 142, 146
nerve, 46, 94
Netherlands, 71
neutral, 50, 62
next generation, 114
NH_2, 50
nickel, 32
nitrogen, 56, 61
NMR, 73
nodules, 15
Norway, 123
notochord, 46

nuclei, 78
nucleic acid, 48, 62, 63, 121, 124
numerical analysis, 165, 168
nutrient(s), 80, 98, 99, 100, 101, 102, 103, 114, 115, 117, 120, 137, 146, 150
nutrition, 126

O

obstruction, 15
OH, 47
Oklahoma, 73
oligosaccharide, 49
operations, 7
optimization, 43, 50, 54, 119, 149, 163, 164, 168, 173, 174
organ, 3, 15, 17, 61, 66, 100, 160, 171, 177
organic polymers, 61
organic solvent(s), 4, 52, 60, 65, 78, 81, 82, 90
organize, 90
organs, 38, 48, 51, 92, 95, 114, 155, 171
ox, 11, 13, 17, 99
oxygen, 8, 11, 15, 35, 98, 99, 101, 103, 114, 171

P

paediatric patients, 119
pain, 15
palpitations, 15
parallel, 94, 176
pathology, 1, 7
pathways, 52
penicillin, 132, 133
peptide(s), 50, 58, 59, 60, 65, 122, 141, 146, 169
perforation, 124, 130, 135
perfusion, 101, 102, 114, 115, 117, 142
pericardium, 31, 32, 169, 170
periodontal, 65
periodontal disease, 65
peripheral blood, 148
permeability, 164, 167
permission, 13, 18
permit, 1, 11, 93, 135
pH, 50, 55, 59, 65, 103, 114, 126, 141
pharmaceutical(s), 54, 63, 66
PHB, 68
phenotype(s), 12, 125, 134, 142, 157, 177
phosphate(s), 72, 73, 84, 88, 122
phosphorus, 61, 62
physical characteristics, 128
physical chemistry, 48

physical properties, 44, 63, 67, 68, 114, 137, 149, 162, 164, 169, 172, 174
physical structure, 49, 97
physicochemical characteristics, 62
physicochemical properties, 46
Physiological, 176
physiology, 22
pilot study, 70
plants, 45
plasticity, 48, 51
plasticizer, 64
platform, 86, 174
polyacrylamide, 73
polycarbonate(s), 68, 150
polycondensation, 56, 59
polycondensation process, 56
polyesters, 52, 55, 150
polyether, 68
polyhydroxybutyrate, 68
polymer blends, 74
polymer chain, 62
polymer composite material, 177
polymer materials, 43, 45, 86
polymer networks, 56
polymer structure, 55
polymeric materials, 1, 161, 172
polymerization, 52, 55, 56, 61, 138, 149
polymers, 4, 38, 44, 50, 52, 53, 54, 55, 56, 57, 58, 59, 60, 61, 62, 63, 65, 66, 67, 68, 69, 72, 78, 80, 82, 84, 85, 86, 88, 89, 96, 120, 150, 169, 170
polypeptide, 45, 46
polysaccharide(s), 47, 48, 49, 50, 73, 89, 92, 136, 159
polystyrene, 136
polyurethane, 27, 58, 73, 142, 143, 144, 169, 170
polyurethanes, 36, 58, 65, 69
polyvinyl alcohol, 89, 92, 95
poor performance, 28
population, 1, 4, 6, 25, 26, 140, 173
porosity, 44, 54, 67, 68, 78, 80, 81, 82, 83, 87, 88, 93, 120, 137, 140, 151, 164, 167, 169
porous media, 176, 177
precipitation, 94
pregnancy, 162, 178
preparation, 41
preservation, 2, 30, 124, 125, 130, 149
pressure gradient, 18, 20
prevention, 29, 159
principles, 98, 173, 174
probe, 72
procurement, 41
progenitor cells, 148, 151, 152, 155, 156
prolapse, 17

Index

proliferation, 3, 4, 51, 58, 65, 67, 68, 99, 127, 132, 134, 136, 137, 140, 141, 143, 145, 146, 147, 149, 150, 151, 153, 157, 169, 170, 171
proline, 47
propane, 55
propylene, 55, 56, 64, 71, 72, 74
prostheses, 2, 33, 34, 41, 154
prosthesis, 29, 30, 31, 34, 35, 40, 41, 42, 175
protease inhibitors, 149
protection, 131
proteinase, 149
proteins, 45, 48, 121, 136, 148, 149, 162, 175
proteoglycans, 12, 134, 158
prototype(s), 83, 87, 89, 96, 143, 169
Puerto Rico, 40
pulmonary circulation, 107, 177
pumps, 8, 114
PVA, 89, 92, 137, 141, 145
PVC, 64
pyrolysis, 66

Q

quality of life, 25
quantification, 40, 142
questioning, 35

R

radiation, 20
radiopaque, 20
radius, 91
raw materials, 68
reaction rate, 136
reactions, 54, 62, 63, 125, 135, 139, 149, 169, 173
real time, 110
reality, 161, 162
receptors, 67
recognition, 44
recommendations, 5
reconstruction, 174
recovery, 50
recurrence, 50
red blood cells, 35
refractive index, 69
regenerate, 97, 122, 126, 128
regeneration, 3, 43, 51, 65, 66, 67, 82, 97, 99, 115, 120, 122, 123, 126, 135, 142, 150, 151, 163, 173
regression, 34, 35
rejection, 30, 136, 163
relaxation, 14, 27, 91
reliability, 26, 38

remodelling, 4, 17, 43, 119, 142, 171
repair, 1, 8, 17, 26, 43, 54, 64, 70, 82
requirements, ix, 4, 26, 67, 68, 77, 99, 161, 164, 166, 169, 170, 172
researchers, 4, 34, 44, 50, 55, 57, 62, 97, 99, 104, 115, 125, 128, 137, 144, 148, 149, 167, 169, 171
residues, 45, 47
resilience, 36
resins, 89
resistance, 25, 27, 88, 105, 127, 131, 167
resolution, 20, 21, 84, 85, 86, 88, 89, 93
resources, 25, 74
respiratory rate, 99
response, 55, 56, 57, 64, 68, 92, 95, 128, 141, 150, 169, 175
restenosis, 64
restoration, 159
retrovirus, 125, 156
rheumatic fever, 1, 7, 8, 20
right atrium, 8, 10, 13
right ventricle, 9, 10, 11, 13, 14
risk(s), 2, 97, 112, 113, 119, 149, 156, 161
rods, 52
room temperature, 59, 65
root, 17, 33, 34, 35, 40, 125, 135, 147, 166
roots, ix, 39, 51
rotations, 114
roughness, 178
routes, 62
Royal Society, 39
rubber, 105, 112

S

safety, 4
scanning electron microscopy, 151
scarcity, 30, 162
science, 69, 73, 154, 174
scope, 168
second generation, 28
secretion, 158
seed, 92, 123, 129, 137, 138, 148
seeding, 4, 82, 89, 93, 97, 98, 107, 122, 123, 125, 128, 129, 130, 131, 132, 133, 136, 138, 139, 142, 145, 146, 147, 154, 158, 160, 171, 172
self-assembly, 90
semilunar valve, 9, 10, 14
sensitivity, 53, 59, 64, 65, 91
serum, 51, 152
sex, 8
shape, 15, 17, 29, 35, 48, 58, 63, 78, 81, 84, 85, 88, 90, 102, 107, 112, 119, 120, 135, 145, 148, 150, 165, 169, 172, 173

188   Index

shear, 27, 28, 35, 37, 40, 41, 98, 99, 100, 104, 106, 109, 112, 126, 137, 161, 164, 166, 173, 174, 175
sheep, 123, 124, 125, 129, 140, 148, 152, 154, 176
shock, 122
shortage, 25, 97
side chain, 141, 145
signals, 44, 97, 112, 117, 173, 176
silica, 74
silicon, 143
silk, 74
simulation(s), 4-6, 105, 115, 166, 171, 173, 174, 177
sintering, 84, 93, 96
sinuses, 17, 33, 137, 143, 165
skeleton, 17, 80
skin, 43, 45, 51, 60, 61, 65, 66, 96, 99, 104, 138
SLA, 86, 170
smooth muscle, 12, 13, 134, 135, 138, 140, 141, 146, 153, 170
smooth muscle cells, 12, 13, 135, 138, 141, 146, 153, 170
smoothness, 86
sodium, 47, 71, 122, 123, 124, 125, 126, 127, 129, 131, 132
sodium dodecyl sulfate (SDS), 127, 132
software, 143, 165
solid matrix, 167
solidification, 85, 90
solubility, 46, 99
solution, 32, 38, 50, 63, 78, 80, 81, 83, 90, 91, 92, 93, 94, 123, 125, 132, 133, 135
solvents, 52, 80, 81, 82, 83, 93
space shuttle, 103
species, 48, 116, 156, 171
specific surface, 67, 149, 170
specifications, 35, 169
spin, 114
sponge, 51, 134, 138, 149, 159
SS, 178
stability, 3, 83, 92, 126, 127
staphylococci, 15
starch, 16, 64, 94
state, 66, 94, 100, 155
steel, 143
stem cell differentiation, 117
stem cells, 4, 75, 132, 148, 155, 160, 161, 171, 172, 174
stenosis, 1, 8, 15, 18, 19, 20, 21, 32, 36, 38, 143, 145, 147
stent, 32, 145
steroids, 153
stimulation, 100, 106, 107, 112, 113, 116, 152, 153, 171
stimulus, 121

storage, 64
strategy use, 3
streptococci, 15
stress, 1, 10, 19, 27, 28, 35, 37, 40, 54, 98, 99, 100, 106, 107, 109, 112, 113, 126, 128, 133, 137, 147, 151, 161, 164, 168, 173, 174, 175
stress factors, 54
stress response, 151
stress test, 151
stress testing, 151
stretching, 91, 108
stroke, 7, 15, 17, 107, 110
stroke volume, 17, 107, 110
stromal cells, 146, 151, 157
structural characteristics, 2, 170
structural protein, 129, 162
substitutes, ix, 25, 26, 34, 39, 65, 119, 161
substitutions, 51
substrate(s), 85, 89, 134, 136, 139
sulfate, 121, 131, 134
Sun, 39, 166, 177
superior vena cava, 13
suppression, 162
surface area, 12, 79, 87, 91, 169
surface chemistry, 77, 120
surface layer, 59
surface properties, 67
surface structure, 164
surface tension, 91
surgical intervention, 162
survival, 18, 39, 41, 171, 172
susceptibility, 158
sustainability, 58, 126
suture, 27, 37, 169
swelling, 15, 60, 65, 176
symptoms, 15, 16, 18
synchronize, 43
synthesis, 55, 56, 59, 60, 62, 70, 90, 116, 135, 137, 150, 152, 157, 175
synthetic polymeric materials, 44
synthetic polymers, 44, 50, 52, 67, 68, 69, 70, 80

T

TDI, 58
techniques, ix, 3, 5, 8, 20, 21, 25, 30, 33, 35, 36, 53, 62, 77, 78, 80, 81, 82, 83, 84, 87, 90, 93, 94, 95, 122, 123, 124, 126, 127, 138, 139, 141, 142, 147, 148, 149, 150, 152, 163, 164, 166, 168, 170, 172
technology(s), 4, 36, 68, 78, 83, 85, 94, 123, 155, 156, 167, 170
temperature, 58, 74, 78, 80, 82, 85, 90, 94, 103, 114, 124, 126

tendon(s), 45, 46, 135
tensile strength, 45, 54, 64, 127, 145
tension, 100, 109, 131, 173
testing, 89, 106, 117, 128, 141, 151, 170
textiles, 70
texture, 94, 125, 141, 164, 169, 170
TGF, 127
therapy, 2, 30, 35, 38, 161, 162
thermal degradation, 54, 70
thermal energy, 80
thermal properties, 59, 65
thermal stability, 66
thin films, 53
thrombosis, 2, 36, 161
thrombus, 28, 35, 37, 148
time frame, 162
tissue engineering heart valves (TEHVs), 3
tissue remodelling, 119
titanium, 27, 29, 37
toluene, 58
total artificial hearts (TAH), 1
toxicity, 54, 58, 63, 64, 65, 98
toxicology, 117
trade, 89
transforming growth factor, 134
transmission, 125, 156
transplant, 30, 32, 38, 119
transplantation, 153, 176
transport, 99, 103, 114, 137, 164
transthoracic echocardiography, 22
trauma, 1
treatment, 1, 8, 25, 30, 35, 65, 124, 125, 126, 127, 130, 131, 149, 170, 174
trial, 39, 44, 92
tricuspid valve, 9
trypsin, 122, 123, 125, 126, 127, 131, 132, 133, 144, 149
turbulence, 29, 37, 101
turbulent flows, 166
tyrosine, 60, 68, 70

U

ultrastructure, 160
ultraviolet irradiation, 46
umbilical cord, 144, 145, 148, 151, 156
uniform, 92, 93, 98, 100, 102, 103, 114
United, 7, 128
United States, 7, 128
updating, 167
urethane, 68, 70, 75, 142
urine, 54

UV, 86, 138
UV light, 86, 138

V

vacuum, 56, 78, 81, 112
valsalva, 17, 18, 143, 165
valvular heart disease, 6, 7, 15, 20, 25, 174
variables, 166
variations, 91, 112, 125, 126, 137, 164
vascularization, 92, 115
vehicles, 61, 66
vein, 124, 129, 132, 144, 148, 151, 171
velocity, 35, 167
ventricle, 8, 10, 11, 14, 17, 27, 37, 106
ventricular assist devices (VAD), 1
ventricular septum, 10
versatility, 52, 59, 169
vessels, 21, 99, 148
viscera, 74
visco-elastic properties, 48
viscosity, 69, 91
vitamins, 51, 70

W

waste, 98, 99, 102, 103, 150
water, 46, 47, 48, 50, 51, 53, 54, 60, 63, 65, 68, 81, 92
water absorption, 46, 47, 50, 51, 53
weakness, 15
wealth, 150
welding, 145, 146
wells, 105, 112
wettability, 66
working groups, 68
worldwide, 1, 7, 25, 26, 30, 36, 68
wound healing, 73, 92

X

xenografts, 119, 156
XPS, 72
x-rays, 20

Y

yeast, 49
yield, 27, 93